The Revolution Will Be Digitised

Dispatches from the Information War

Heather Brooke

D0299824

✳ WINDMILL BOOKS

Published by Windmill Books 2012

2 4 6 8 10 9 7 5 3 1

First published in Great Britain in 2011 by
William Heinemann

Windmill Books
Random House, 20 Vauxhall Bridge Road,
London SW1V 2SA

www.randomhouse.co.uk

Addresses for companies within The Random House Group Limited
can be found at: www.randomhouse.co.uk/offices.htm

The Random House Group Limited Reg. No. 954009

A CIP catalogue record for this book
is available from the British Library

ISBN 9780099538080

The Random House Group Limited supports The Forest Stewardship Council
(FSC®), the leading international forest certification organisation. Our books
carrying the FSC label are printed on FSC® certified paper. FSC is the only
forest certification scheme endorsed by the leading environmental
organisations, including Greenpeace. Our paper procurement policy
can be found at: www.randomhouse.co.uk/environment

Typeset in Dante MT by Palimpsest Book Production Limited,
Falkirk, Stirlingshire
Printed and bound by CPI Group (UK) Ltd, Croydon, CR0 4YY

He who receives an idea from me, receives instruction himself without lessening mine; as he who lights his taper at mine, receives light without darkening me. That ideas should freely spread from one to another over the globe, for the moral and mutual instruction of man, and improvement of his condition, seems to have been peculiarly and benevolently designed by nature, when she made them, like fire, expansible over all space, without lessening their density in any point, and like the air in which we breathe, move, and have our physical being, incapable of confinement or exclusive appropriation.

Thomas Jefferson

Contents

Foreword ix

Chapter 1: Data in the Desert 1
Contingency Operating Station, Hammer, Iraq,
January 2010

Chapter 2: Building a Revolution,
One Hackerspace at a Time 16
Boston, Massachusetts, Wednesday 27 January, 5.45 p.m.

Chapter 3: Turning Science Fiction into Fact 34
Reykjavik, Iceland, Thursday 25 February

Chapter 4: Welcome to Wiki Wonderland 56
Tønsberg, Norway, Saturday 20 March

Chapter 5: In the Belly of the Beast 76
National Press Club, Washington DC, Monday 5 April

Chapter 6: Land of the Free? 93
Cambridge, Massachusetts, June

Chapter 7: Private Lives 123
London, July

Chapter 8: The Information War Begins 167
Berlin, October

Chapter 9: To the Brink and Beyond   199
Cambridge, Massachusetts, 21 to 24 November

Conclusion: A Brave New World   227

Afterword to the paperback edition   241

Acknowledgements   251

Index   253

Foreword

We are at an extraordinary moment in human history: never before has the possibility of true democracy been so close to realisation. As the cost of publishing and duplication has dropped to near zero, a truly free press, and a truly informed public, becomes a reality. A new Information Enlightenment is dawning where knowledge flows freely, beyond national boundaries. Technology is breaking down traditional social barriers of status, class, power, wealth and geography, replacing them with an ethos of collaboration and transparency. In this new Enlightenment it isn't just scientific truths that are the goal, but discovering truths about the way we live, about politics and power.

During the first Enlightenment the free flow of information was considered essential to understanding the natural world; without full disclosure we had no hope of overcoming our inherent human biases that occluded our vision of the truth. In England, scientists were careful to cordon off this questing curiosity to science but its revolutionary impact in politics led to the American and French revolutions. Thomas Jefferson said that America was an experiment that would 'demonstrate to the world the falsehood that freedom of [speech and] the press are incompatible with orderly government'. America produced 'the first legislature that had the courage to declare that its citizens may be trusted with the formation of their own opinions'.

This aspiration is not solely American. Citizens around the world have long declared a desire to be trusted with the formation of their own opinions, and that can only come when they have access to the facts. This is the essence of the information war. Do we trust citizens to communicate freely and come to their own conclusions, or do we believe those in authority have a right to restrict and manipulate what we know? Do we hold to Enlightenment ideals of reason and the pursuit of truth no matter where that takes us, or put our faith in authority to make certain an uncertain world?

The digital age can be experienced as a sensation of information overload, a deluge of data and accelerated technological innovation. As I became enmeshed in this world I felt overloaded, too, as I experienced the erasure of boundaries between the public and private that characterises the online sphere, and found myself grappling with the slippery nature of digital data first-hand as data leaked and then the leaks were leaked. It can seem like chaos. When anyone can say anything how do we determine what is true and what is misinformation or lies? Exposing secrets can and does reveal corruption and abuses of power. Powerful figures who once relied on deference to keep silent those who would speak against them can now be challenged by weaker individuals who have a global voice thanks to the Internet. Such a publishing free-for-all can also leave individuals vulnerable to false claims.

Additionally as we move our lives online we are in danger of creating a central storehouse where nosy state officials and private companies can monitor our every move. Where are the boundaries between transparency and privacy; quality control and freedom? We have the technology to build a new type of democracy but equally we might create a new type of totalitarianism. These are the stakes in the information war.

The Internet is powerful because it allows people to organise around issues at unprecedented speed, broadcast their thoughts and challenge those in charge. A wave of such groups banded together in early 2011 to demand the removal of authoritarian leaders in the Middle East as one country after another rose up with varying degrees of success. But the Internet doesn't cause revolution. It is a communications network. What people choose to *do* with technology – that is where we can make moral judgements. Some people will use it for ill, others for good. Security forces tend to focus on the ills, while the majority use it for good. In the name of protecting us from 'bad things on the Internet' there are increasing moves to suppress communications networks in both repressive and democratic countries. Demands to shut down, censor, filter or in other ways oversee and control the way people communicate are on the rise.

The extent of the digital revolution was brought home to me as I was putting the final touches to *The Silent State*, an account of my investigations into Britain and the power imbalance between citizens and the state. For five years I'd investigated British institutions mostly using the country's new Freedom of Information Act. I was born and raised in America, and trained as a journalist there, so maybe this was my way of 'bringing the revolution home' by trying to open up secretive British institutions to the citizenry. I made many hundreds of FOI requests to various public bodies but the one that provoked the greatest resistance was to Parliament, which fought for five years – ultimately all the way to the High Court – to prevent publication of how parliamentarians were spending public money.

At every stage of my investigation politicians and bureaucrats were convinced they could avoid accountability as they always had before: first by calls to deference (we had to trust them), when that failed, obstruction (this is none of your business), followed by misinformation (claiming publication would be dangerous and harmful) and finally outright refusal. Even after I won the court case, officials worked to suppress and stifle the information. What no one predicted was quite how easy it would be to simply copy the entire digital database of politicians' expenses claims and leak it to the public. In 2009 someone inside Parliament did just that, and the resulting full disclosure caused a major political scandal.

What we witnessed in Britain was a culture clash: bewigged and be-stockinged parliamentary officials who believed in their right to rule without public prying; suddenly challenged by a populace no longer content with this arrangement and newly armed with an entire data set of how politicians were spending public money. Officials tried to shut down and manage the story, but it was precisely their heavy-handedness that fanned the flames of public discontent. One politician after another was forced to resign.

In this book I chart the battles fought over information globally. Two sides are competing to determine the world's future: on one side freedom fundamentalists, hacktivists and democracy campaigners seeking to increase the free flow of information largely through the Internet; opposing them are the traditional gatekeepers of information – governments, police, intelligence agencies and the military.

I explore the most pressing issues of the digital age: digitisation and why it is revolutionary, hackers and hackerspaces, the law in a globalised world, the role of journalism, information ownership, privacy, anonymity, Internet surveillance and national security. The narrative traces the year of my investigation, and while I could discuss the information war without mentioning WikiLeaks, the events of 2010 were so dramatic it seemed natural to use them as the spine of the book and nestle within each chapter wider discussion of the relevant issue. What I hope to show is that this revolution is not simply about WikiLeaks. It's just as much about what is happening in Iceland, where citizens rose up to overthrow official secrecy, or the hundreds of young Internet activists setting up spaces around the world where they come together in an online version of the English coffee houses where the Enlightenment was born. Times of great change can feel like chaos, and the natural reaction is to crave certainty. Instead of trying to impose control, what I suggest in the conclusion of this book is that we look afresh at the way we organise power. I believe there is a better way to practice politics and the Internet provides a model.

A few notes on the text. To convey the drama of events I have written the book in a narrative style. Some scenes were pieced together, particularly the first chapter, which is a portrait of the leaker, here described as an anonymous 'kid', who I have created based on information in the public record. Whether or not Private Bradley Manning is guilty of the leaks, he carried the consequences for most of 2010 and 2011 and so I have taken some of his biographical details to describe this character. Chapter Two is reconstructed from interviews with David House and Jim Stone. Other chapters written in novelistic fashion are based on interviews with the following people: in Iceland: Birgitta Jónsdóttir, Smári McCarthy and Herbert Snorrason; in Boston: David House, Danny Clark and a few others who did not want to be named; at the *Guardian*: David Leigh, Alan Rusbridger, Nick Davies and Ian Katz. Others who contributed are mentioned by name in the text but I also took information for narrative scenes from interviews with Daniel Domscheit-Berg; Rop Gonggrijp, James Ball, Jacob Appelbaum and Julian Assange. In other chapters, I relate events and conversations as I experienced them first-hand.

1

Data in the Desert

Contingency Operating Station, Hammer,
Iraq, January 2010

The funny thing about life here: you never really know the time. It could be noon, dusk, dawn or midnight, though it hardly matters. Once you punch the five digits into the cipher lock you enter a world where time is meaningless, like the dark heart of a Vegas casino devoid of clocks. *Welcome to Casino America*. Hey, why not? It is, after all, a cheap construct of America plopped down in the middle of a desert. The guys outside could be playing basketball in Ohio or Texas; only their tan camos and sand-coloured boots give any hint that this is a war zone. The Pop-Tarts, Tater Tots and Cheerios are just like at home. It's just a shame so few Iraqi families are going to be enjoying such delights.

The kid hangs up his jacket and checks out the half-full Mr Coffee on the counter. Another day stretches in front of him. It could be a day just like the one before, a day that he can hardly recall as it was so similar to the one before that, like reflections in a mirror. It's virtual reality, and the irony is that life online feels more real. That's where his friends are and where he feels most connected. But today is going

1

to be different. Not that the other guys in the room will notice, he hopes. One of the young men looks up from his computer. He smiles but his eyes quickly zip back to his screen.

Another man in a khaki T-shirt nods a curt acknowledgement. He doesn't really like this kid. Maybe it's because he's so short, or the way he's so serious and silent, as if he's superior or something. Fortunately there's no need to make a lot of small talk as the first thing the kid does once he settles at his desk is put on his headphones. Maybe he's listening to music or watching videos or whatever. The kid is an intelligence analyst so there's all sorts of material to keep him busy.

The room is barren and beige, the makeshift walls decorated only with a few whiteboards, a poster of the Stars and Stripes and a world map; a television hangs from the ceiling in one corner, a chunky white printer sits idle in another.

The kid pours coffee into a chipped, oversized US Army mug and heads to his workstation. Unlike the other desks in this prefab building, his area is neither spartan nor plastered with sports memorabilia. There's an American flag poking out of a pen pot and what look to be a couple of photographs tacked onto a wall. Neither of the other guys has bothered to ask who the people in them are, and the kid has not volunteered. He logs onto his computers and the light from the screens flashes into his clear blue eyes. He's only just out of his teens with apple cheeks and an open expression somewhat tempered by a solemn, quiet demeanour. Some part of him would like to be less serious, see this place as a joke like the others, 'just get out of it what you can'.

Partly that was the reason he'd joined the army: to get

some qualifications and experience, give his life focus, direction. He'd always been good with computers and did a few classes at community college but . . . how could he describe it? The world has so much to offer and he couldn't wait to get out and explore it. He did want to be all he could be, and working meagre jobs to pay for community college wasn't it. The military had worked out for his dad and so his dad thought it would do the same for the kid. Early on, he'd thought maybe his dad was right. Technology was making all sorts of things possible but society wasn't changing fast enough. Being part of the US Army would help spread the gospel of democracy, Internet freedom and freedom of speech throughout the world.

He scrolls through his emails. There are a few requests from his commanding officer. Some admin work that he processes quickly. He logs into the classified networks using his username and password. On one computer he logs onto Siprnet (the Secret Internet Protocol Router Network) and on the other, JWICS (the Joint Worldwide Intelligence Communication System), the top-secret version of the Internet used by the State Department and Department of Defense to transmit classified information. He puts on his headphones and pops in a CD marked 'Lady Gaga'. The rules are that devices like CDs or memory sticks are only supposed to be used for making official copies or backups, and iPods are banned, but you only have to look around to see how strictly these rules are enforced, which is to say not at all. Everyone on the base is watching videos, movies, listening to music, playing games. How else are you supposed to keep yourself sane for fourteen hours a day, every day, with no breaks and little else to do? The kid was surprised at first to see guys putting music CDs into machines hooked up to classified networks, but after three weeks at the base you stopped caring.

Any ideals he'd had about the invincibility and security of the American military networks were demolished after his first week. He'd asked the guy from the National Security Agency if he'd ever seen any suspicious activity coming out of the local networks but all that had got him was a shrug and a muttered 'It's not a priority,' before the guy went back to watching the film *Eagle Eye* (presumably preferring the cinematic version of his job to the mundane reality).

But the kid took his duties seriously. His job description stated he was to assist 'in determining significance and reliability of incoming information' and that's exactly what he did, digging through multiple intelligence reports of the same incidents, getting independent verification where possible or a translation if needed to find the truth. That's what he'd done when his commander asked him to investigate the detention by the Iraqi Federal Police of fifteen people accused of printing 'anti-Iraqi literature'. The Iraqi police wouldn't cooperate with US forces, so the kid was told to find out who the supposed 'bad guys' were and the significance of their crime. He got hold of the material and found an interpreter to read it to him. There was nothing in the literature advocating violence or supporting terrorists; rather it was a scholarly critique against the Iraqi Prime Minister Maliki titled 'Where did the money go?' It was exactly the sort of First Amendment freedom of expression that the US was supposed to be exporting.

He went immediately to the officer in charge, and set about explaining how the suspected terrorists had simply written a benign report on political corruption that you might see in any newspaper in a democratic country. It was a mistake, he insisted. These guys were allies, not enemies – they were targeting the wrong people.

This wasn't what the officer wanted to hear. It didn't fit into his game plan, his worldview, and his eyes began to drift long before the kid had finished his little speech.

'I didn't ask for your opinion on the matter,' the officer replied, icily. 'I want you to get their names, their addresses. And more where they came from.'

'But they are not our enemies.'

'You don't know that,' the officer shot back. 'They could be.'

'But they're not. And by concentrating on them we're ignoring the real enemies.'

All this had done was infuriate the officer, who ordered him to mind his own business and do the job he'd been tasked to do.

The kid pounded his fingers onto the keyboard. *If this is the calibre of people charged with making Americans secure then goddamn, we are in some serious shit.* It was as clear as the sun shining down every day that if the US military carried on like this then Al Qaeda wouldn't need any other recruiting agent than the good ole United States Army.

The officer's disparagement of the truth led the kid to wonder what other lies were being told about Iraqis and to the American people for the sake of some official's easy life. He'd channelled his frustration into his intelligence role by trawling through the military networks looking for files or incidents that showed similar cover-ups. There were a lot of codes and tags and jargon but he could piece some of it together from what he already knew. He found a cache of military dispatches from the current war in Afghanistan.

Key	4B54C6BD-3057-4C3B-8C74-30342C5D9670
Date	2008-11-19 17:10:00
Type	Friendly Action
Category	Escalation of Force
Tracking no.	20081119171041RPQ2750094200
Title	(FRIENDLY ACTION) ESCALATION OF FORCE RPT J COY 42 CDO : 1 CIV KIA
Summary	***DELAYED REPORT*
	J COY 42 CDO conducting a admin move to MOB LKG observed 1 x LN vehicle approaching the convoy and fired 2 x Miniflares as a warning. The LN vehicle did not change course or slow down so FF fired 2 x RDS 9mm (WARNING SHOTS) into the ground in front of the vehicle. It swerved and stopped. A visual check of the LN vehicle carried out and the patrol continued on task. Subsequent investigation revealed that a LN child was present in the vehicle and received a fatal GSW. The incident was reported to NDS by HAJJI HAQBIN (An influential BARAKZAI tribal elder) a relative of the child.
	***Event closed at 261808D*1 Killed None(None) Local Civilian

There were even more files from the Iraq war, nearly half a million of them. He wasn't sure what truth these records would reveal but he suspected it was different from the 'We're winning' whitewash that the Pentagon was churning out to the American people.

Even if the commanding officer didn't appreciate his thoroughness, the kid thought other soldiers and operatives in the field would. It was a bit like playing God – watching over all these people. He wasn't trying to control them, just making sure they were safe and that they got the most accurate picture of reality so they could make the best decisions. Maybe that's what made him think about leaking the data.

After all, the stuff on Siprnet wasn't exactly top secret. How could it be, when all that was required to access the database was a security clearance up to 'Secret' which some 3 million US soldiers, government employees and contractors had? A democracy, we're told, is government by the people, for the people, so why should the common people be kept in the dark? They're the ones paying the bills and in whose name wars are fought. Was it right that this information be restricted only to those who already held power? *This is America as all Americans should see it. Power talking to itself.*

Whatever his reasoning, today is a turning point. From helpless voyeur of history to change-maker. He is feeling strong today with the courage of his convictions. It's no good looking to anyone else to make the world a better place. We each have to do our part to work with what we're given.

The more he thought about it, the more amazing it was that no one had done it before. A corporation couldn't get away with the sort of lax security that existed on the base. They would have been poked and prodded by competitors. But what competition does a government have except other governments? It was about 100 per cent likely, he figured, that China and/or Russia already had everything worth taking from this system. Scoured through the files, sucking down data quietly and covertly. He'd mentioned a few obvious fixes early on, but no one was interested in fixing anything. More work. As far as he could tell, there was only one other person who knew anything about computer security on base, a SIGINT (Signals Intelligence) analyst, but fixing flaws mustn't have been his job either. *You see where the bureaucrats' priorities are. For them national security is about securing their own positions of power.*

So why hadn't anyone made this information public before? His only conclusion was that people are trained to do what they're told, rarely stopping to consider whether it's right or wrong. Someone in authority says 'Do this,' or 'Don't do that,' and most people will blindly follow, especially in the military. Weren't we past that whole Nuremberg defence thing yet? he wondered.

That was the appeal of WikiLeaks with their motto, 'Courage is contagious.' Watching what they had done had certainly changed the kid's views on what was possible. The publication of the 9/11 pager messages in November, messages that were strictly classified and held by the National Security Agency, had impressed him. You didn't get hold of them without having an inside source and you didn't publish without having balls of brass. And that, apparently, was what WikiLeaks had. They championed the public's right to know in the face of trenchant official secrecy.

He'd also been reading about Daniel Ellsberg, the military analyst who leaked the Pentagon Papers to the *New York Times* during the Vietnam war. When the idea first occurred to him, he couldn't imagine placing himself alongside a man like

Ellsberg. But then as he began digging into the data, he realised that what he read day in, day out could be today's Pentagon Papers.

War could be noble if you were in it for the right reasons and acting honourably. But the longer he was here, the more he believed this war met neither of those qualifications. His privileged access to the inner workings of the US and other governments had also led him to another, even more startling revelation. There was a bigger war going on. An Information War. And in this war, the kid wasn't just an errand boy. By teaming up with WikiLeaks it was possible he could transform global perceptions about politics, power and people's right to know. He asked himself if his reasons for entering this war were noble and just. He was fighting to liberate not only the American people, but the world's people, from rulers who wielded power over their populaces not just through authoritarianism and fear, but through ignorance.

People who change history need to be prepared to make sacrifices. Words are easily spoken but the ultimate test must be action: what are you prepared to do? What consequences are you prepared to endure to stand up for what you believe? If he was serious about bringing truth to the world, changing power structures, liberating information and giving power to the people then it was time to contact that crazy, white-haired Australian. He could go to a journalist but which one? And would they even publish such material? He doubted it.

Inspired, he carefully starts to search through the directories looking for the items he'd mentally catalogued earlier. His plan: to transfer the data from the classified network onto a disk, erase all evidence of the download, then get the data onto his other laptop where he'll sort it, compress it, encrypt it and upload it to WikiLeaks. There are a few details to work through and he has a trip home planned at the end of the month, a well-earned break. Maybe he can seek advice. But for now, he can at least make a start.

He synchs the words to 'Telephone' while writing the war logs to disk. 'Poker Face' might have been more appropriate.

Digital data

Before we go any further I want to define 'digital data' so we can better understand why it is so revolutionary.

Before digitisation, data was stored in paper, microfiche or tape format. Digital comes from the Latin *digitus*, meaning 'toe' or 'finger'. 'Digits' came to refer to numerals because the ten fingers (*digita*) used to count correspond to our decimal system. 'Digital' is defined as anything that is relating to, resembling or possessing a digit or digits, and digital data is data represented as a series of numerical values or information displayed as numbers. What is remarkable about digital data is that its duplication and transmission can seemingly be done beyond any physical manifestation. It has an ethereal quality, almost like pure thought. Ether used to be an imagined substance that was believed to fill all space, and support the transmission of electromagnetic waves. 'Ethereal' means something is light as air, impalpable, celestial or spiritual. The huge troves of data that exist on Siprnet or JWICS have this quality. The data exists, but where? And how? Not in physical stacks of paper, but somewhere inside a network. The physical component is the server or servers on which bits of the data are stored but in its transmission to another server it has no physical mass.

What this means is that digital data can be shared and spread easily, just like an idea or a thought. It can replicate so rapidly that its proliferation can often be equated to the spread of a virus. The cost of duplicating digital data is zero, or close to it, once you have the necessary hardware to store it on. Think back to the analogue age when it cost money to share information as it meant producing another physical copy, whether a printed book or a cassette tape. In the digital age, these physical components are gone. Now it costs money *not* to share data. Sharing and copying are the default dynamic of digital data and any person or organisation who wants to impede this free-flow has to spend considerable amounts of money and resources. Unfortunately for the information liberators of the world, there are plenty of individuals and organisations willing to outlay the necessary resources to do this, as we will read later.

By 'digitisation' I mean the transfer of data into digital form so it can be processed directly by a computer. It's this ability to directly interrogate and analyse information that is so powerful and revolutionary. When I teach journalism students about what is quaintly called 'computer-assisted reporting'[1] I tell them to study the scene in *All the President's Men* where the actors playing Bob Woodward and Carl Bernstein are sitting in the Library of Congress trawling through thousands of library cards looking for any members of the Republican Party who borrowed books on Ted Kennedy. The way the scene is filmed gives some flavour of the time and the tedium of such research in the analogue age. Now all this data is digitised in a central database and can be interrogated in all manner of ways while still providing an answer within seconds.

1. I take issue with this phrase as we don't talk about 'telephone-assisted reporting'. The fundamentals of investigation remain the same regardless of the technology used.

Which brings me to another aspect of digitisation: the huge amount of information that can be stored on comparatively minute pieces of physical matter. There's more information in the world than ever before – as former Google CEO Eric Schmidt told an audience in August 2010. Every two days we create as much of it as we did from the dawn of civilisation up until 2003. Yet storage has also changed dramatically. The ratio of data to matter on a stone tablet is pretty poor. It gets better as we move to vellum or paper and better still as we shift to magnetic ribbons. It's truly staggering once you get to disks, USB sticks and microchips. Two major technological advances – in semiconductor manufacturing and fibre-optic communications – have made it possible to cram a lot more information into smaller physical components. The price performance of silicon has been accelerating so that a computer that cost you $1,000 two years ago would cost just $500 today and would be twice as powerful. This exponential rise is known as Moore's Law after Intel co-founder Gordon E. Moore, who described the trend in his 1965 paper; it has proven uncannily accurate ever since. Now a stick the size of a thumbnail can easily contain a database of 250,000 records which in their paper form, just to put this in perspective, would work out to some 213,969 pages of A4 paper that would stack about 25 metres high – not something one could easily slip past security.

The contrasts with the analogue age couldn't be starker. When Daniel Ellsberg sought to enlighten the US public about the government's secret activities in Vietnam, he had to spend hours in front of a photocopier to duplicate the Pentagon Papers, then physically hand the material over to a newspaper reporter. When a court injunction stopped publication in one newspaper, he had to give another hard copy to another newspaper, and then another. US Senator Mike Gravel then read portions of it into the Congressional Record. With so many copies around, the secrecy became untenable. When the White House challenged the legality of this hugely controversial and politically damaging leak, the US Supreme Court ruled that the unauthorised release of the Pentagon Papers was necessary and that only a free and unrestrained press can effectively expose deception in government. What's new today isn't just that one database can hold six years' worth of sensitive military data, or that hundreds of thousands of people have access to this data and can download copies, but that an upstart website like WikiLeaks can upload just one of these copies and share it with the entire world using a series of uncensorable, and untraceable, international servers hooked into a global network. If you're an authoritarian government partial to controlling your citizens through controlling information, that's a frightening prospect.

Yet the very ephemerality of digital data also presents a danger. If pages are ripped from a book there is a gap in pagination, so we spot it. But when virtual matter is altered, how do we know? When British newspapers receive libel threats over articles, for example, it is not uncommon for the paper to instantly pull the article from its website. If the reader knows the link to the article, she is met with a '404 error' which indicates only that the server could not find the requested page, which might be for any number of reasons. It's as if the article never existed. This was one of the difficulties UK campaigners faced when trying to reform the libel law in 2010/11 because they couldn't get accurate data on how often newspapers were self-censoring after receiving a libel threat. Similarly in China, this is how news which is online at one hour suddenly disappears the next without a trace.

Digitised data can leave an audit trail, particularly if a decent records management system is in place, but often it takes a computer forensics specialist to trace it. People may not be aware that pressing 'delete' on a computer to get rid of something doesn't instantly make that item disappear. It remains on the computer – or mobile phone – until it is overwritten. It is rather like a letter written by hand in pencil. The words are erased but there may still be traces of lead, or indentations from the pressure of the handwriting, from which words can be deduced. Computer forensics experts seek to retrieve deleted digital data in a similar way. There are ways to counter this – one of which is to write over the disk, a process known as zero-filling where each byte of data is imprinted with a zero or sometimes a '1'.

While governments usually have the resources to track down and reconstruct deleted data, most of us don't. If we buy a printed book, for example, and then it is recalled, the store, publisher or government will have to physically take the book from us, which once you've sold even a few hundred copies is no easy task. Not so with digital. On 17 July 2009, thousands of people who had downloaded 99-cent copies of George Orwell's *Animal Farm* and *1984* to their Kindle eBook readers found the titles suddenly deleted. It transpired that Amazon.com remotely deleted the titles from purchasers' devices after discovering the publisher lacked the proper rights. Amazon refunded customers' money but many were furious. Some likened it to a bookstore clerk coming to your house when you're not at home and taking the books from your bookshelf.

It's fitting that one of the books dropped into this memory hole was *1984*. In that novel, Orwell describes the largest section of the Ministry of Truth's records office as the part employing people 'whose duty it was to track down and collect all copies of books, newspapers, and other documents which had been superseded and were due for destruction'. This was essential if all the past versions of news were to be altered or deleted to maintain the 'truth' of the present. And herein lies one of the gravest threats of digitisation. In the digital age, this most difficult part of any censorship operation – that of collecting, altering or destroying the offending speech – becomes easy. Information stored digitally on a centralised network can easily become one giant memory hole, and the way authoritarian governments rewrite history is itself being rewritten.

In the digital age we have the technological tools for a new type of democracy but the same technology can also be used for a new type of totalitarianism. What happens in the next ten years is going to define the future of democracy for the next century and beyond.

2

Building a Revolution, One Hackerspace at a Time

Boston, Massachusetts,
Wednesday 27 January, 5.45 p.m.

It's fifteen minutes before the launch and the basement of 111 Cummington Street is filling up. David House, a lean twenty-two-year-old with blond hair, Nordic cheekbones and a passing resemblance to the lead singer of eighties Norwegian band A-ha, is making final preparations. He has on a top hat so people can identify him as the ringleader and a black T-shirt with the words 'GRAFFITI RESEARCH LAB' printed in white. At 6 p.m. the Boston University Information Lab & Design Space (BUILDS) is officially open.

About sixty people file into a room that resembles a set from a low-budget science fiction show. On one side there are banks of PCs and servers, on another a mural of a 2001 *Space Odyssey*-style floor-to-ceiling computer and a porthole showing a view from space. Whiteboards, chalkboards, toolboxes and power tools fill the room. There are several tables, one laden with soft drinks, another with tiny coloured lights, another strewn with metal hooks and picks.

'What is this place?' a woman asks the man next to her, gazing around.

'How can I get a swipe card?' a student asks David.

'Just follow me,' he tells the crowd in his soft East Coast accent. He was born and raised in Alabama, but he's long since lost his Southern drawl. He trained himself by watching old Humphrey Bogart films but if he drinks enough whiskey the Southern accent creeps back.

'Can I get one too?' Soon everyone wants a swipe card. It gives access to the room 24/7, the only one of its kind anywhere at BU. Down the road at the prestigious Massachusetts Institute of Technology, workshops like this aren't unusual, and the students there are deemed trustworthy enough to have access at all hours. Until now, though, the more sober BU admins haven't been so keen to entrust their students with their own room filled with power tools and computer servers. David has spent the past few months talking them round, persuading them that a research centre created and run by students, with an entirely open and fluid agenda, will be a benefit to the university. It will help foster a spirit of innovation and entrepreneurship needed to create the tech companies of the future. A video crew mingles among the crowd and picks out a member of staff to interview. She thinks the space is a great idea, as do a number of her colleagues.

It didn't start out that way.

When David arrived as an undergrad at BU in January 2008 eager to soak up as much knowledge and as many new ideas as he could, there wasn't an active computing club in the computer science department. He'd spent years back home walking in the woods, dreaming about a future surrounded by smart people making their dreams happen, and yet BU was proving too conventional. The local chapter of the Association for Computing Machinery (ACM), the world's largest computing society, had lapsed, and no one had bothered to renew it. Plus it seemed that the ambitions of the majority of his fellow computer science students reached only about as high as getting a job at an insurance company. Conventionality was exactly why he'd left home, where he'd always felt out of place. He'd managed to get to Boston on the strength of his intelligence and while it wasn't MIT, he hoped to bring some aspects of MIT to his university. David decided to set up a local ACM chapter (though he didn't register it with the national chapter – 'We just kind of adopted the name,' he says). His goal early on was to achieve legitimacy, get professors on board, enlist local support and revitalise the practically non-existent computing community at Boston University.

One of the group's first projects was to test the security of the new BU student ID/debit card, as David suspected it wasn't as secure as the university administrators claimed. In September 2009, he and a few others discovered the cards could quite easily be 'hacked' using reverse engineering (whereby careful analysis of the finished product reveals the underlying data set) and that students' social security numbers and other personal information could be harvested as a result. David is quick to point out that 'as soon as we found this flaw we started talking about it'. The group's discovery made it into the student newspaper and as a result David received a 2 a.m. call from the university's general counsel strongly advising him to quit the project.

The ACM students backed off, the security errors they'd highlighted were fixed, and they carried on holding their meetings, which were by now quite popular, only at the next meeting a man with a badge turned up and sat in the room taking notes. He was Jim Stone, whose official title was the Director of Consulting Services, but he was also charged with investigating security issues on campus. He'd been sent by the administration to keep an eye on the troublesome students. 'The university was a little concerned to say the least,' Stone told me later. 'They'd heard this rumour about a hacker group decoding the ID cards and the worry was that they would then start to create fake cards for fraudulent purposes. I was very upfront with them and they were equally upfront with me. They told me they were curious about the system and making fraudulent cards was not their intention.'

Stone came to all their meetings and listened as various speakers talked about phone tapping, lockpicking and computer security. Stone had never seen students talking about these things before, and he was impressed. The students in turn gradually started thinking of him not as an investigator but as a friend, and one with a shared interest in computer security. Stone reported back to the university that the kids were all right, and in a strange twist, the man sent to spy on the students became their patron. It turned out Stone was from an influential BU family that already had a library and science museum named after it. Jim decided to donate $10,000 to the ACM.

The group sat on the money for a while, wondering what to do with it, then had an idea. It was clear that what they needed was their own space, so they began petitioning the department of computer science for a deserted, leaky old room they'd found in the basement. They managed to strike a deal with the administration: let them use the room for a while, and in return they'd fix it up. The students then asked if they could have twenty-four-hour access to the room with their own keys, and 'modify the walls a little bit'. There was some hesitation, but eventually the administrators agreed. The entire building was going to be torn down in fifteen years anyway.

The ACM now had everything they needed legally to set up

a space, so they took Jim Stone's donation and set about transforming the room. They did most of the work at night, when no one was in the building. David would bring in power tools and he and the other members would drimmel out the door and install an electric strike for card access. They put in networked computers, plastered and painted the walls and brought in artists to paint murals.

The administration hadn't expected that. A few posters maybe, but not wholesale refurbishment. They didn't know what was happening until a few months later, in late December 2009.

'A professor came down and saw this crazy hippy hang-out in the middle of the university,' David says. 'The professors were split on how to feel about it. The ones who had come from MIT thought, "This is awesome." But the administration, the secretaries – we were a huge headache for some of them. Here's this club that found a load of problems in the ID card system and now they have a permanent base of operations in the school.'

What's more, this base had autonomy. While the university specifies that school clubs can't have a space unless they meet certain requirements, BUILDS was not itself a club, only a project of the ACM, meaning that when problems arose the students would direct them back to the ACM, bypassing the administration altogether. It was a sort of legal hack of the school rules. The project then became very political and David responded in kind. ACM members joined committees that dealt with IT. One of the guys dressed up in a suit and went over to the office of the dean of students every other day to make friends with him. That worked and they proceeded to get other campus officials onside.

BUILDS has no constitution and exists because of the support of high-profile donors and encouraging officials within the university. That is how it wants to exist – because this isn't just about technology; it's about creating an environment of empowerment. David sees it as playing a crucial role in the lives of the university's students.

'You take kids at a university who were doing nothing. Who were going about their classes and following the status quo and all of a sudden you bring them into a room and you tell them they can rearrange it, paint the walls and cut down the ceiling if they want to. Then you show them that locks are psychological barriers. You tell them the social system they live in is a system like any other and they can hack it if they want to. You're empowering them in every sense of the word. You're telling them that they, themselves, can be leaders in the world simply by their actions. That's an incredibly powerful message to give to someone and it's exactly the message that a good university should be giving to its students.'

Hackerspaces

Although BUILDS is officially a 'research centre' it could just as well (and perhaps more accurately) be called a hackerspace. It has many of the features of hackerspaces around the world: computers, power tools, tech books, electronic art. There's a red robot computer mouse; the hooks and picks are used by the Open Organisation of Lockpickers to give a demo on the practicalities and legalities of picking locks. A representative from Graffiti Research Lab shows students how to make LED 'throwies' – simple magnetised LEDs that can be thrown on buildings to create luminous graffiti art.

Hackerspaces range in size from a garden shed to an industrial warehouse but the occupants usually share a love of learning and a dislike for anything that gets in the way of that learning. They are places where tech people go to tinker on projects either real or virtual, individually or with others, in person or across the Internet – they're like open community labs or workshops where people come together to share knowledge and build things. I've been to tiny hackerspaces like HACK (the Hungarian Autonomous Center for Knowledge) in Budapest where members have built an electronic plant-watering system, and Sprout in Cambridge, MA, where an MIT student was building a jet propeller; and to large spaces like Noisebridge in San Francisco where members created an active space exploration program sending weather-balloon probes up to 70,000 feet in the sky to collect images and data using GPS smartphones and digital cameras. There are new hackerspaces like BUILDS popping up all over the world, and older ones like the German Chaos Computer Club and C-base, both based in Berlin, that have been around for decades.

Hackerspaces are the digital age equivalent of English Enlightenment coffee houses. Much like them they are places open to all, indifferent to social status and where ideas and knowledge hold primary value. In seventeenth-century England, the social equality and meritocracy of coffee houses was so deeply troubling to those in power that King Charles II tried to suppress them for being 'places where the disaffected met, and spread scandalous reports concerning the conduct of His Majesty and his Ministers'. It was in the coffee houses that information previously held in secret and by elites was shared with an emerging middle class. They were social but also educational and political places. French author Antoine François Prévost visited English coffee houses at the time and described them as the 'seats of English liberty' where any man had the 'right to read all the papers for and against the government'. It was this freedom of information that led to the creation of

Lloyd's of London and many other businesses that fuelled the British Empire. Publications such as the *Spectator* became integral to coffee-house life and some ascribe the magazine as being one of the main instruments that led to the vast social reforms of the eighteenth century when English public life was transformed.

But what precisely *is* a hacker? It's worth giving this a bit of thought as the ideals of the digital age are very much tied up with hacking culture, so if you're imagining teen boys locked in their bedrooms trying to steal credit card details from an online database then you'll have the wrong idea.

Hackers describe what they do as playfully creative problem solving. They believe it's much easier to attack than to defend a system. This is because there are many more people trying to break into a system than defend it. Hackers value ingenuity and intelligence above all and one needs more of both to build something than to break it. The best hackers, therefore, are those who build things. Crackers are the ones who break them. The World Wide Web, and free software operating systems like the GNU Project and the Linux kernel, could all be considered hacker creations. Even Facebook is a hacker creation.

That is not to say hackers don't take things apart. They do, with a compulsive glee. Hackers want to deconstruct systems to figure out how they work. It's the difference between taking apart a car engine to see how it operates and ripping it apart purely for the thrill of destruction. There are some hackers who do the latter but the most respected have a compulsive need to know how things work in order to build the most effective technological solutions to problems. Even this definition is too simplified. Some of the best hackers say the line between hacker/cracker or white hat/black hat (e.g. good/bad) is of little relevance. What is more important is the ethics behind the hacking. A hacker can use his skill to protect a system he knows is used to track down and kill protestors. He's not 'cracking' but how can he be considered a white-hat hacker? Another hacker might break into this very same system and damage it but by doing so save people. Does that make him a black-hat cracker? The ethics of hacking, like life, are slightly more complicated than a 1950s Western movie.

Most of what hackers believe can be boiled down into four basic principles:

1. Freedom of information: information is a resource that does not diminish with sharing, rather it increases. It is therefore wrong and counterproductive to the social good for information to be treated as a limited commodity such as land.

2. Meritocracy of ideas: the best idea wins. You shouldn't have to solve the same problem twice, which isn't to say that ideas can't be improved upon but that solutions shouldn't be hindered artificially through conventions of status, power or greed.

3. Joy of learning and knowledge: everyone should have the opportunity to learn and contribute with few restrictions on the information necessary for such learning.

4. Anti-authoritarianism: someone can be an authority on a subject, which is fine, but authority over another human being simply because of power differences is not fine. This type of 'because I say so' authority goes against all the previous principles, stifling access to information, accepting poor ideas for the sake of status, etc.

Hackerspaces aren't just about hacking with computers, though that is certainly their humming heart – they also contain power tools, industrial cutting machines, sewing machines and sometimes even kitchens for 'culinary hacking'. These principles equally apply to philosophy, sociology and politics. Politics is, after all, a system by which humans live and work together, so like any other system it can be 'hacked'. In this sense the existence and operation of hackerspaces is a real-time experiment in political hacking. It is where the principles can be put into practice.

Access to and membership of the spaces is usually governed by commonly agreed norms, but what is notable about most hackerspaces is their lack of formal rules. The idea of asking permission from someone is frowned upon as it implies a power imbalance between the person asking for permission and the one giving it, too much like a parent/child relationship, seen as one of the main problems with the way society is currently organised.

'People are actually afraid of freedom. Afraid of *not* having to ask permission,' says 27-year-old Jacob Appelbaum, one of the founders of Noisebridge, the California-based hackerspace that developed its own space program. He is one of the world's prominent hackers (or independent computer security researchers, if you prefer). 'When people come to Noisebridge for the first time,' he claims, 'and want to, say, paint a mural on a wall, they ask everyone if they can paint a mural. A month later. They see. They've observed. They know the answer. They ask around. But they don't ask for permission, rather for preference. They paint the mural because they realise they have the authority to do so. I think a lot of people are not ready for that. People are conditioned into a state of fear when it comes to expressing themselves or their desires.'

When I visit Noisebridge, sure enough I just walk in. No one asks who I am or what I'm up to. There's a guy without a shirt brewing some kind of weird tea called kombucha in the kitchen. He's friendly in that laid-back Californian way (even though I discover he's actually from the Czech Republic). He offers me a cup. There are plenty of worn sofas for hanging out; computers, power tools, posters and strings of coloured lights hanging from steel rafters. There's an area of shelves filled with programming books and on the wall above, a poster from the movie *Bill & Ted's Excellent Adventure*. It reminds me of my college newspaper office but with a hippy, New Age vibe. This impression is cemented when (another) shirtless guy comes over and kisses my hand and introduces himself as Salvador Dali Lama. It's not all geeks in black T-shirts and hoodies in the hacker community.

The *Bill & Ted* poster, I find out later, expresses Noisebridge's governing principle: Be excellent to each other.

I catch up with Appelbaum in Seattle, where he's now a staff research scientist at the University of Washington. We meet first for a coffee and then he takes me to his favourite Thai vegan restaurant where we continue what will become a day-long conversation about everything from computer networks, politics and power, to Buddhism, veganism and the Church of Satan (he practises all of the last three). Even in the world of hackers and Web revolutionaries, his dedication to the Internet is startling (it is 'the only reason I'm alive today', he told *Rolling Stone* magazine). He found in computers an escape from a chaotic upbringing. He was very close to his father, who had several jobs including acting but was also a heroin addict. Jacob once came home from school to find a suicide note signed by his father. He saved his dad that time, but a few years later he died in what the police called a drug overdose but what Jacob calls murder. He says his mother, whom he lived with for his first five years, is a paranoid schizophrenic. He then lived with his aunt for two years until she dropped him off at a children's home. That's where he says he hacked his first security system at the age of eight, getting a PIN code from a security keypad. He credits his skill in technology for moving him out of America's underclass and into the middle class. As well as working at the university he's also a spokesman for Tor, a free Internet anonymising software that helps people defend themselves against surveillance, and he's spent five years teaching activists how to install and use the service to avoid being monitored by repressive governments. Now he himself is frequently targeted as a result of his friendship with Julian Assange and the fact that WikiLeaks uses the Tor software.

Appelbaum founded Noisebridge with two others because he wanted a place to hack that wasn't a coffee shop where he had to pay money to hang out.

'What we've done at Noisebridge is not to say how bad everything is but to create a viable alternative. I wanted a space where we could make things come true. Almost like a magical environment where we could decide one day we wanted to have a space program and then – we did. That's not going to happen in a café. There currently isn't a public place where you can have a lathe or a table saw or computer access or couches where you can sit, where no one feels entitled to throw you out. That's what is different about a hackerspace. The closest thing is the university lab where I work now but there we're beholden to university administrators. In this place we are beholden to no one but ourselves.'

I ask Jacob how the place is run. What *is* the political system? He says the political ideology of a hackerspace is probably nearest to libertarian 'in the liberty sense': anyone who wants to contribute something, whether time, money or ideas, is welcome. The ability to do things is premised on being accepted by the group and that comes from the merit of one's actions. Appelbaum goes on to say that Noisebridge is run along anarchist principles. Mutual aid. Solidarity. Respect. He's keen to stress that this is not anarchy in the sense of chaos but as author Emma Goldman describes it:

> Anarchism, then, really stands for the liberation of the human mind from the dominion of religion; the liberation of the human body from the dominion of property; liberation from the shackles and restraint of government. Anarchism stands for a social order based on the free grouping of individuals for the purpose of producing real social wealth; an order that will guarantee to every human being free access to the earth and full enjoyment of the necessities of life, according to individual desires, tastes, and inclinations.

OK, but what about the practicalities, I ask – how do you make decisions? Spend money? Hold meetings? Do you use some procedural rule book to run a meeting?

'Someone mentioned *Robert's Rules of Order* as a good example of how to run a meeting. I said, "I have a better idea. How about we just talk with each other."'

Have you ever been to a local council or school board meeting? I venture. Because I've been to plenty and frankly the idea of having to listen to everyone in the room speak without limit is enough to send me running for the door.

Appelbaum admits that's the downside of listening to what everyone has to say: 'It can take a while.' But the advantages outweigh the costs, he believes. 'In our society people don't have a lot of agency to change things. And when I say "our society" I mean "the world". Sometimes in our discussions it will be the first time that someone has ever listened to a person in their entire lives. That's actually incredibly sad but I'm happy we can give them that opportunity. I think there are a lot of us who don't think the world is as we would like to see it. And anyone who is not a utopianist is a schmuck.'

There must be a leadership structure though. Even a computer directory has a hierarchy.

Appelbaum explains a concept he calls 'sudo leadership'. It comes from the term 'substitute user do'. On operating systems like Unix with multiple users the superuser or root user is the one with the highest privileges; a goal of many hackers is to become the superuser so they have total control of the system. The command 'substitute user do' allows a regular user to perform a command as the superuser, so sudo leadership is about letting people take the lead on things in which they have an interest but without being the leader for ever.

I admire his hope for a better future and the way he's modelling out his own principles in practice. But I wonder if the way one runs a hackerspace can actually be scaled up to governments. If a Stalin-type joins a small group like Noisebridge then it's easy for the group to discover that fact and reject him. But once the group goes beyond a certain size then consensus-based systems just don't work. Appelbaum counters: 'When I hear people say that Noisebridge wouldn't scale to a government, I think that shows a failure of imagination. Democracy doesn't scale either. What we call democracy here is not what the Greeks would call democracy; how many redistricting and zoning issues do you suppose they had? Only monarchy really scales and even that isn't effective if you look at historical trends. But this doesn't mean you shouldn't try. Until you try you just don't know.'

Fair point, although some of the worst genocides in history have resulted from people trying out new political ideologies on their populations.

Appelbaum's mention of ideological systems 'scaling' got me thinking about how niche groups can sometimes punch above their weight. There are subsections in society everywhere in the world but for various reasons the values of one sometimes gain an influence that is disproportionate to the group's size. In Saudi Arabia, a puritanical and extremist sect of Islam has become vastly more influential than its number, due to its geography sitting on top of a vast oil supply. The wealth that has poured into Saudi Arabia has enabled the practisers of Wahhabism to fund global missionary activities and overwhelm less puritanical interpretations of Islam. Saudis have spent some $87 billion propagating Wahhabism abroad during the past two decades.

The hacker community may be small in number but it sits atop the technologies that are driving the global economies of the future. It is no surprise that as we have shifted from an industrial to a knowledge economy, those who build the infrastructure and products of this economy have made their fortunes and found their values moving into the mainstream.

Geek was once a term of disparagement; now it's used to describe intelligence, ingenuity and potential millionaire status. The hackers of the 1970s and '80s founded billion-dollar companies such as Sun Microsystems, Cisco, Microsoft and Apple and, having made their fortunes, often used their wealth to promote the hacker ethos. Later in the book I'll talk to John Gilmore, who made his fortune at Sun Microsystems and went on to use his money to help create the free software movement and co-founded the Electronic Frontier Foundation to defend digital freedom.

Sometimes, though, hackers don't realise they're no longer outsiders but have their hands on the levers of power. Tech writer Danny O'Brien points to Bill Gates, who continued to believe right up until the anti-trust lawsuits of the 1990s that Microsoft would be destroyed by big powerful companies like IBM – despite the fact that he was, by then, one of the world's richest men and Microsoft the world's biggest IT company. Furthermore, it can prove difficult to reconcile the subversive hacker mindset with the demands of running a multinational firm. A robust disregard for intellectual property in youth, for example, is often replaced with a cadre of lawyers enforcing draconian copyright law once the requisite information has been hacked, the product built and the company made profitable.

O'Brien used to write a tech newsletter in the nineties called *NTK* (*Need to Know*) and mused about starting it up again. But the times they have definitely changed. Sipping cocktails with him in an upscale Palo Alto restaurant, I can't help noticing that this once grubby Londoner is himself an embodiment of the transformation. 'When we wrote it originally, geeks were under-represented, under-discussed; an unconnected group of outsiders who were excited about technology. But now the story is how much ridiculous power we have and how we're misusing that power or unaware of it.'

Was David House aware of the new power of geeks? Maybe so, because he certainly wasn't nervous about how the university would respond to the launch of the first hackerspace in its midst.

'I can't wait to see what happens. I can't wait to see how people react. Will they be enthralled or pissed off or angry? Maybe the police will bust things up. Shut us down.'

Doesn't that make him worried?

He says no, he's not worried. 'This is what university is about, it's what life's about. Doing crazy things and seeing what happens. If they succeed that's awesome, if they don't, try something different later. If it messes up then we'll change it and try again. I'm a person of action. Not to the extent that I do things without thinking. I consider things but it's important to act.'

The three weeks preceding the launch of BUILDS have been manic. David and a few other students have worked night and day to get it ready: repairing leaks, painting, letting the artists loose on those walls, hooking up the computers. Right until a few hours before opening they've been scrambling around.

Now it's come together: the students are enthusiastic and there's a good showing from members of staff. David, in a rare moment alone, surveys the scene. The students around the lockpickers' demonstration are smiling. He hopes they're beginning to understand. Lockpicking is the art of overcoming a system whose sole purpose is to prevent your success; escaping this system is difficult and requires a person to push mental boundaries and improve understanding through direct experience. Lockpicking, he thinks, is a symbolic manifestation of the hacker's relationship with the world. To pick a lock for the first time is to gain an understanding that all barriers in your environment, all limits, are purely psychological. You hold the key to your own liberation.

As he's scanning the room, a short guy of about his age is doing the same. This guy looks impressed. Their lines of sight cross. The visitor wanders over and sticks out his hand.

'Hey, my name's Brad.'

'I'm David.'

'Cool. Did you create this space?'

'Yeah,' responds the host, with a glint of pride. 'Have you been to a hackerspace before?'

Brad nods his head but it's not clear if he means yes or no.

'Are you from Boston?' David asks after a moment.

'No. Oklahoma.'

'Cool. Who do you know here?'

'A friend of mine lives here.' Another pause. 'I think I've seen you at Pika House before.'

'Cool.' David thinks he remembers the guy, but there are always visitors at the shared house and with twenty or so tenants it's hard to keep track of who is a visitor and who is living there. It's mostly MIT people and their friends, girlfriends, boyfriends, randoms. 'Well, nice to meet you,' he says, the conversation apparently run its course. 'Hope to see you around.'

'Yeah. Good luck with the space. It's a great idea.'

'Thanks.'

As he turns to go, David can see the guys from the lock-picking table calling him over, and he has some more swipe cards to make. If he's thinking anything just then it's that he wants to meet everyone in the room before the night is over.

3

Turning Science Fiction into Fact

Reykjavik, Iceland, Thursday 25 February

A North Atlantic gale rushes past wharves and warehouses, swirling into the centre of Reykjavik, the capital city of Iceland. Austurvöllur Square is empty, the flower beds frozen, the bare basalt walls of Parliament House dusted with snow. It's the dead of winter when night tips into day, spilling out shadows. But this evening, from the upper windows of the two-storey stone building, light radiates. It gives shape to strange reliefs carved into the rock sills: a bird, a bull, a giant and a dragon, Iceland's guardian spirits.

Although it might look like a city hall, Parliament House is where the Althingi (Parliament) has met since 1881 when it moved from its previous headquarters inland. That site was getting a little worn, having been in existence since AD 930 when the Old Icelandic Commonwealth was created. Inside Parliament House is a hall of simple Scandinavian elegance, the grey stone plastered smooth, a sunny yellow in the entranceway; a seascape of blues and greens in the main debating chamber. The ceiling is studded with gold-orbed chandeliers. During the parliamentary session, the public enter the building through a modern annex off to the side. The security is not overbearing. Icelanders tend to vent their rage against their leaders not with rifles but by rifling through their kitchen cupboards for pots and pans which they bang loudly in protest marches under the windows of Parliament House. The most recent people's uprising was thus called the kitchen revolution and it has transformed politics in Iceland. If the bill under discussion tonight goes through, it may do the same for the world.

There are no sounds of saucepans on this freezing February evening, only the howl of oceanic gales. Inside, three men sit in the public gallery not in the least concerned with the weather. They are focused on the Member of Parliament below who has risen to speak on a project close to their hearts.

Birgitta Jónsdóttir MP speaks the rough-hewn Icelandic words that will lay the foundation for what will become known as the Icelandic Modern Media Initiative.

'Our population stands at a crossroads and changes to the law are needed,' she tells the nearly empty debating chamber. 'The proposals in this report will transform our country, creating and strengthening democracy. They are a catalyst for necessary domestic reforms: for greater transparency and accountability.'

Whether the handful of MPs in the chamber share her dream of making Iceland the world's future information haven is unclear. They look at their order papers or the clock while the forty-three-year-old MP speaks with a clipped, somewhat dry intonation. Behind the rather utilitarian delivery, however, lurks the most radical shake-up of freedom of speech laws since the American Revolution. The resolution, if passed, will mark the first real attempt to take an idea from science fiction and make it a practical reality.

Jónsdóttir is an unusual sort of politician. She describes herself as a poet but before she was elected to Parliament she made her living running websites and coordinating technology projects. A Tibetan flag hangs from the rear-view mirror of her small car. Her mother was a well-known singer in Iceland, and her father died when she was young – wandering off one day, he was found a year later, dead in a river. She lived in Australia for several years and then Norway before resettling with her family in her home country. Her eyes are deep ocean blue, tinged with sorrow but full of conviction. Her long black hair is braided and pinned up with a peacock feather, a severe fringe cut just above her eyebrows.

She lays out the basics of the resolution. If passed it will pave the way for thirteen new laws on all aspects of free speech: from libel and privacy to freedom of information and protection of sources. They are not completely new laws, rather they are based on existing laws from trusted jurisdictions: the press freedom law from Sweden, the New York state law on libel, a source protection law from Belgium. In the digital age, when information is borderless, countries need a holistic approach to freedom of expression to protect against well-funded vested interests censoring publications. And that, Birgitta intimates, is precisely what happened in Iceland. If the Icelandic people are ever to trust their banks, politicians or media again, this resolution must go through, she says. The time for secrecy and unaccountable power is over.

'It is hard to imagine the amazing resurrection of our country from financial ruin and widespread corruption due to secrecy, but we intend to offer a business model based on transparency and justice. By taking legislation that has proven its worth around the world, we aim to create a comprehensive package of laws that address all aspects of modern media and information freedom. We will be first in the world to market ourselves as a country with a principled, holistic and modern set of laws fit for the digital age.'

Her words echo around the carved wooden benches.

With only a few people here tonight it might seem her heartfelt words are wasted, but speaking for the resolution is a formality. The real work will be getting it through committee and successfully voted upon in the full chamber. Even so, there is at least one person in the audience who will appreciate her advocacy: Julian Assange, the founder of whistle-blowing website WikiLeaks.

To understand why Julian Assange was in Iceland we need to go back to the financial collapse of the country's economy, a crisis largely considered to have been helped along (if not caused) by excessive secrecy. Iceland in 2010 was ripe for reform, like the USA just after Watergate.

Iceland's history is replete with get-rich-quick schemes, but this penchant for boom and bust reached epic proportions when newly privatised banks and official secrecy mixed with the global credit economy. Credit was easy and Icelandic banks loaned huge amounts of money. One businessman hired Elton John to sing 'Candle in the Wind' at his fiftieth birthday party. Ordinary Icelanders took out mortgages secured on the full value of their homes in dollars, euros and yen. They bought Land Rovers and fancy sports cars on credit. In his book *Meltdown Iceland, Times* reporter Roger Boyes estimated that the country's 'financial elite' comprised no more than thirty people, with the newly privatised banks funding the politicians who had privatised them and a revolving door between finance, politics and the media. If it all seemed too good to be true, in late 2008 it turned out it was.

The Icelandic banks were making very large loans to their own shareholders, sometimes more than the bank's entire capital. They were giving very favourable terms to those borrowers, and not checking if they were creditworthy. Effectively, the banks were investing in themselves: they were lending money to people whose major assets were shares in the bank. This in effect reduced the capital of the bank, so their capital ratio was actually much higher than they claimed, and much riskier than their credit rating implied. The credit didn't match the reality and in October 2008 the country went bankrupt. Once a status symbol, Land Rovers were renamed 'Game Overs' as interest rates rocketed to 18 per cent after the country had to take a loan from the IMF. The Icelanders were saddled with massive debts to pay back the UK and other countries whose investments were held in Icelandic banks. More than 50,000 people had their savings wiped out. Boyes described the mood in 2009: 'Icelanders were conscious that they had been steered incompetently and corruptly into the abyss. Yet the political class showed no remorse.' Initial attempts to investigate how the collapse had occurred and who was responsible came back with either muted silence or finger-pointing. By January 2009, the Icelandic population was angry and the kitchen revolution began. The aim was to bring down the government.

In the wake of the uprising, the Prime Minister resigned and in April 2009 a new government was formed. Birgitta Jónsdóttir had been one of those protesting with pots and pans outside Parliament House and in just eight weeks she went from outside to inside Parliament. Later when I met up with her in her tower-block apartment she told me, 'During crisis times you can actually get real change.' The goal was to revolutionise politics, starting with secrecy. Amidst the black-and-white photographs of her Icelandic ancestors she said that in Iceland people wanted real change. 'Not just a rewriting of banking laws but do we want to have a banking system like this where we can all be conned?'

In August 2009, WikiLeaks got hold of the loan book of one of Iceland's biggest banks, Kaupthing, which revealed the extent of the bank's unusual lending practices. The state television channel RUV tried to report it as their top story, but a few minutes into the news, anchorman Bogi Ágústsson had to say he couldn't actually report the top story because, for the first time in Iceland's history, the channel was served with an injunction, brought by Kaupthing Bank. He was, however, able to announce that the injuncted documents were available on WikiLeaks and he made sure to spell the name out again with a WikiLeaks logo screenshot as background. Icelanders piled onto the Internet and were livid, not only at what they read but at the bank's attempt to stop them reading it. All of a sudden WikiLeaks was famous – at least in Iceland.

An independent 'truth commission' was appointed by Parliament to investigate the financial collapse. Icelanders were angry: at the banks, at the politicians who had failed to act in the public's interests, and at the media for failing to report independently. An indication of the conflation of politics, finance and the media is illustrated by Davíð Oddsson, the head of the Central Bank and one of Iceland's longest-serving politicians who as Prime Minister had presided over the privatisation of Iceland's publicly owned banks. When he was forced to resign, he quickly became editor of the country's leading newspaper, *Morgunblaðið*.

In Iceland, secrecy was now the ultimate sin. Transparency and public accountability were the new watchwords, and this is why Iceland was such fertile ground to plant the seeds of a radical transparency movement.

Jónsdóttir met Julian Assange just three months ago but already his impact has been immense. And not just on her. Flanking Julian in the public gallery are two board members from Iceland's Digital Freedom Society: Smári McCarthy, twenty-six, and Herbert Snorrason, twenty-four. Iceland's foray into radical transparency started when these two young men invited WikiLeaks to speak at the society's annual conference in December 2009. It was Herbert's idea, one that popped into his head just two weeks before the conference. He'd been impressed and inspired by WikiLeaks' role in publishing the Kaupthing documents. Herbert and Smári had been sitting opposite each other in the corner of a 3,000-square-foot former fishing factory that now housed among other start-up ventures the 'Hakkavélin', a hackerspace whose name translates into English as 'the hacking machine'. They sent an invitation and waited. Smári didn't rate their chances but surprisingly, the next day Herbert received an email from someone at WikiLeaks agreeing to attend.

'Great,' said Smári when Herbert told him the good news. 'When are they coming?'

Herbert didn't know when, or indeed who, would be coming, and no further emails were answered. They waited anxiously for the best part of a fortnight, and just two days before their scheduled appearance at the conference, WikiLeaks' second-in-command Daniel Domscheit-Berg (who at that time was using the pseudonym Daniel Schmitt) arrived. Julian Assange came later that night on the last flight into Iceland. Smári and Herbert couldn't believe their luck – these weren't just representatives from WikiLeaks, they were the masterminds behind the whole organisation. They all ended up that night at the guest house where the WikiLeaks guys were staying.

Daniel and Julian were obviously used to living in close quarters and there seemed a natural affinity between the two. They were both focused and intense when they talked about WikiLeaks. Daniel appeared the more technically minded of the two. He was a lean, analytical German in his late thirties with closely cropped dark hair and dark brown eyes peering out intensely from dark-framed rectangular glasses. He wore all black and spoke English in a precise and careful way that belied the boldness of his ideas. Herbert was pleased to discover in Daniel a fellow fan of one of his favourite intellectuals, Pierre-Joseph Proudhon, and they talked about the French philosopher's famous treatise on anarchism, *What is Property?*, which Daniel claimed was the most important book ever written.

Julian, by contrast, was his own philosopher and his ideas came from the creative forge of his imagination, helped along by the novels he'd read. He was taller than Daniel and light where his colleague was dark. When he shambled in with his backpack, Daniel reported to him what he'd already discovered about Iceland. There was an easy relaxed manner between the two men and it was clear they shared the same drive to make WikiLeaks a revolutionising force in tackling official secrecy. Julian was the charismatic leader, and Daniel took on a parent-like role of making sure everything ran smoothly behind the scenes. Daniel had met Julian at the Chaos Computer Congress in Berlin two years previously and thought the idea of an uncensorable leaking site so wonderful that he'd quit his IT job and devoted all his time and money to the venture. Julian had even moved in with him and they worked side by side uploading secret documents from their laptops in Daniel's apartment.

Keen to capitalise on their presence, Herbert arranged for them to go on the country's biggest talk show, where they discussed the Kaupthing Bank scandal they'd exposed. Their appearance caused such a stir that even politicians sported WikiLeaks badges and people in bars bought them drinks. They were treated like heroes.

'A crisis is a terrible thing to waste and Iceland has a lot of opportunity to redefine its standards and legislation,' Julian told the presenter that night. 'Iceland has seen some of the problems that happen when society becomes too secret.'

His solution was for Iceland to become a centre for global publishing by introducing a strong set of free speech and whistle-blower protection laws.

Smári and Herbert were sitting in the production room listening. 'Did he just say that?' They looked at each other. He had. It got them thinking.

At the conference a couple of days later on 1 December, Julian and Daniel expanded on the idea. They had first-hand experience of how difficult it was becoming around the world to publish information of vital public importance. They related the way that English libel law, in particular, was being used globally to stifle reports of corruption and abuse of power. Julian described the case of Trafigura, a multinational commodities company which, in 2006, dumped toxic waste off the Ivory Coast. Afterwards, some 108,000 people ended up in hospital but when the *Guardian* newspaper tried to report on the case they were threatened with a libel action by Trafigura and served with a super-injunction by Carter-Ruck solicitors (a London law firm synonymous with bringing libel actions). The super-injunction prevented the publication of a leaked scientific study commissioned by Trafigura (the Minton Report) that contained damning evidence of the potentially toxic nature of the waste, and also prohibited the *Guardian* from mentioning the injunction itself. A member of the UK House of Commons raised a question in Parliament about the super-injunction in order to get its existence into the public record but when the *Guardian* tried to report on that, Carter-Ruck threatened that it would be a contempt of court. Meanwhile the documents were leaked to foreign media and WikiLeaks in an attempt to thwart the draconian English reporting restrictions, and news of the super-injunction spread virally around the Internet, largely through Twitter. On 13 October 2009, the law firm dropped its demand to the British media to censor the parliamentary record, writing three days later that the newspaper was 'released forthwith' from any reporting restrictions. Not so much a Twitter revolution as a Twitter toxic avenger.

Yet that was just one of some 300 super-injunctions in England alone at any one time according to Index on Censorship. There is also a growing trend by powerful public figures and organisations to use privacy law to shield themselves from public accountability.

Most freedom of information campaigners and investigative journalists find themselves on the defensive, battling to stop the constant incursions powerful people make to limit the public's right to know. There is another strategy, however: go on the offensive. The advantage is that it forces the other side to spend their time and resources defending themselves. By taking a position of radical transparency the campaigner can potentially redefine the boundaries of the playing field (or battlefield if you want to use the information war metaphor). After what had happened in Iceland there was an opportunity not just to defend the status quo, but to propose something revolutionary.

'If one country's laws can be used to stifle speech globally,' Julian concluded, 'then it's equally possible that another country's laws can be used to protect speech globally. Iceland could be that country.'

Data havens

The idea of a free speech or free data haven is not new. It's a common trope in the science fiction of the 1970s, most notably the cyberpunk writings of Bruce Sterling and William Gibson. Another seminal influence on modern-day hackers was John Brunner's novel *The Shockwave Rider*, which came out in 1975, featuring as the hero a rogue computer hacker who frees data for public good (an information Robin Hood, as it were) while being chased by operatives from a dystopian surveillance state.

When privacy legislation was first introduced in Europe, a British government report in 1978 expressed worry that differing privacy laws around the world could lead to 'data havens' and that Britain might become such a haven where data was not 'protected'. Data havens were thus rogue countries who wouldn't sign up to privacy laws, rather like a data version of the 'free ports' from pirate days. The science fiction writers of the time were good at picking up on, and rethinking, these ideas, refashioning the data haven as an information 'free port'.

Like many tech people who came of age in the late seventies and early eighties, IT writer Danny O'Brien, whom we met in the previous chapter, first heard about data havens and hacking from the writings of Sterling, Gibson and Brunner.

'These books were great influences on my generation,' O'Brien says. 'The idea of running off to this pirate utopia – it was common currency in science fiction and cyberpunk writings. The cyberpunk hero is the hacker who somehow manages to negotiate the power structures and use his awesome skills online to lay everything out to the general masses. Certainly most hackers have read these books.' One of the most popular authors in this genre is Neal Stephenson, who in 1999 further developed the idea in his book *Cryptonomicon*, and went into great technical detail about the logistics of creating a safe harbour where information could be shared freely or hidden securely from government interference or censorship.

Data havens have thus lived in the imagination for decades, with few actual attempts at making them a physical reality. One of these came on 2 September 1967, when Major Paddy Roy Bates, a British citizen, took over an old Second World War fort off the coast of Britain with the intention of broadcasting a pirate radio signal from a floating pontoon lodged onto a sandbank about 6 miles off the coast. The distance was crucially important as it put him outside the 3-mile territorial water claim of the United Kingdom. In the UK at that time, the state not only controlled the radio broadcasting spectrum but had a virtual monopoly on the stations themselves through the British Broadcasting Corporation (BBC). Not surprisingly, such central control failed to meet the listening needs of the public and as a result a variety of pirate radio stations started up, operating from anchored ships or marine platforms outside the UK jurisdiction. In the end, Bates didn't broadcast but he did declare Sealand (as he named the fort) a 'sovereign principality' and appointed himself HRH Prince Roy of Sealand. Nothing of any significance happened until 2000 when he launched HavenCo Limited, a data hosting services company that promised to operate as an information 'free port', prohibiting only child pornography, email spamming and malicious hacking, but allowing everything else. Copyright or intellectual property restrictions would be ignored as Sealand was not a member of the World Trade Organization or the World Intellectual Property Organization. The launch was widely reported in the media, even making the cover of *Wired* magazine. Partner Ryan Lackey, described by *Wired* as a twenty-one-year-old MIT dropout and self-taught cryptography expert, thought the idea would make him rich. Then the Twin Towers fell on 9/11, data freedom fell out of fashion and Lackey fell out with the Bates family, leaving in 2002. The idea never really took off either, as Internet traffic still had to be routed via the UK.

In 2007, a Swedish organisation called the Pirate Bay tried to buy Sealand but the Bates family refused their bid. At the time, the Pirate Bay was the world's largest BitTorrent tracker. BitTorrent, invented in 2001, is a system for distributing large digital data files, sharing them over the Internet between an unlimited number of recipients. A BitTorrent tracker is a directory service, much like a card catalogue in a library. It lists all the files that might be available in the system and describes their content and where they can be located. The vast majority of the files listed in the index cards of the Pirate Bay are digital copies of movies, audio recordings and computer software. The Pirate Bay has no role other than acting as a directory; what people choose to share with each other is up to them and the site is rather like Google in that it is a search engine (albeit for torrents). It was originally set up in November 2003 by the Swedish anti-copyright organisation Piratbyrån (the Piracy Bureau) but became independent in October 2004. It has been on the receiving end of numerous legal threats from music, movie and games companies who claim the content of the torrents violates copyright laws, particularly the American Digital Millennium Copyright Act of 1988 which criminalises copyright infringement. The Pirate Bay begged to differ, stating that it was a Swedish company based in Sweden so only Swedish law applied; and under Swedish law it was not violating copyright as it was merely acting as a host for information sharing, not uploading or storing any information itself. It is another instance of one country's law (in this case the US Digital Millennium Copyright Act) being used to suppress information in a global jurisdiction. The owners of Pirate Bay were eventually prosecuted and found guilty of violating Swedish copyright law after the law and its enforcement was changed (some say due to intensive political lobbying by American government officials[1]). They are currently appealing their case.

Laws are constantly changing but in the area of technology they are not evolving nearly fast enough to keep pace with a new kind of global nervous system that lets people communicate between disparate parts of the world.

Smári and Herbert were familiar with data havens. They'd read *Cryptonomicon* and while the idea of turning Iceland into a transparency jurisdiction wasn't quite as glamorous as setting up a pirate's free port, it was still pretty damn cool. Nor were Julian and Daniel the first people to propose the idea. At the

1. US diplomatic cables revealed that the government offered more than $500,000 to fund a recording industry-backed copyright enforcement initiative in New Zealand. American officials also lobbied several cabinet members while New Zealand was working through its copyright reform in 2008. The cables show similar lobbying efforts in Canada, Sweden and Spain.

previous year's Digital Freedom Society conference, John Barlow, a former songwriter for the Grateful Dead and co-founder of the Electronic Frontier Foundation, and Eben Moglen, a law professor at Columbia University who set up the Software Freedom Law Center, pitched the idea of Iceland becoming the 'Switzerland of bits': meaning that just as Switzerland attracts business by promising a jurisdiction of banking secrecy, so Iceland could create a market by offering a jurisdiction of digital transparency. The US and other governments' expansion of Internet surveillance would create a market opportunity to provide companies with a place where they could base their data centres without having to become de facto agents of the state, hassled to intercept, collect and hand over their customers' personal information. Herbert and Smári were keen on the idea but they had no clue how to implement it. Until they met the guys from WikiLeaks.

Birgitta Jónsdóttir was also at the conference on 1 December and as she listened to Daniel and Julian detail their battles to publish important public news, she realised this was exactly the sort of project Iceland needed and she was in the perfect position to help make it a legislative reality. She was an MP, she was an activist and she had a technology background. Later that evening, she organised a meeting at Peace House, a hall in Reykjavik, to discuss the idea further with Julian, Daniel, local activists and journalists.

'OK,' Birgitta began, getting straight to the point, 'can you please give us an idea how something like this could actually work?'

'Think of how a tax haven operates,' Julian said to the assembled audience of mostly young people in their twenties. (Birgitta, at forty-three, was likely the oldest.) 'Now, what Iceland could do is follow this model but instead create an information haven. It would become *the* global publishing portal – because soon all publishing will be moving to the Internet.'

The current piecemeal legal approach to regulating information was unsuitable for the way it now spreads globally, he continued. Some law firms, such as Schillings in London, had discovered they could use one country's laws, namely England's libel law, to litigate globally. But WikiLeaks based servers in, or routed data through, various countries so that if challenged it could take advantage of the most protective laws, such as Sweden's press laws, Belgium's law on source protection and the United States' First Amendment.

The activists listened as Daniel and Julian gave one example after another of battles fought and won to publish information in the public interest. There were the Church of Scientology manuals over which they'd been threatened with copyright violation, and political pressure from the German and Australian governments after they'd published Internet blacklists that showed censorship of all kinds of non-criminal websites. Suddenly there seemed a real, tangible solution to Iceland's problems, and the activists saw clearly for the first time how the only thing holding back political reform was their own limited conviction that they could make a difference. The message from WikiLeaks was that Iceland needed to change, and it would only take a few committed activists, particularly when they had technological skill and political currency, to change society in a profound way.

It was 10 p.m. before they finished. The core group found the only restaurant still open – a tapas place tucked along an alley near the waterfront – and filed in, all talking at once. Among the terracotta tiles, Birgitta, Smári, Julian, Daniel, Herbert and a couple of other Icelanders couldn't stop discussing all their plans to reform society. Birgitta was fascinated by the idea of Iceland becoming a data haven and at one point she turned to Julian, her eyes searching his.

'Let's do this. I mean, really do this.'

Smári flew to Berlin after Christmas for the annual Chaos Communication Congress where Daniel and Julian were giving another talk about WikiLeaks to a crowd of supportive hackers. He stayed in the city for two weeks, immersing himself in free speech law. Julian's list of laws had come from memory, but Smári set about reading the laws in full, writing them down and finding more.

Smári, in common with most of the information activists I've met, and a good many computer programmers, has a flexible working schedule. He makes money by occasionally selling software or doing contract programming but he also allows himself plenty of time to work on various projects he finds interesting, usually in the areas of information access and digital freedom. He loves travelling and in the upcoming months has trips planned to the Netherlands, Italy and Sweden where he'll meet in person many of the activists he talks to online. When I first met him his reddish blond curls were tucked under a bowler hat. He speaks with an Irish brogue and appears mild-mannered until a few things slip out such as, 'If you take a fish bomb and put it next to the wall of a house it has a lot of sonic energy. It doesn't do any real damage but it does shake the house so everyone screams.'

He lives in a house near Birgitta on one of the main roads, about five minutes' walk from downtown Reykjavik. It's a house shared with four other twenty-somethings. Some are students, one woman is over for a few months from Sweden, another is a mathematics graduate who is thinking about moving to Berlin. Many evenings turn into impromptu gatherings where the housemates, their friends and friends of friends gather to talk about mathematics, bands, computing, travel, democracy, art. Someone may put on a CD (the Swedish electronica band The Knife is a current favourite) and before you know it everyone is dancing and doing shots of the appropriately named Icelandic liqueur Black Death. On the bookshelf there's Neal Stephenson's *Cryptonomicon* (of course).

Iceland's demographics show it was settled by people of Nordic and Gaelic origin, with a genetic study indicating that the men were Norse while the women were Gaelic. You'll hear differing theories in Iceland about why this is so: that a boatload of strapping Viking men landed in Ireland and the women lusted after them to the point of asking to be carried off, or that the Vikings raided towns and kidnapped the women. Smári's ancestry is the opposite: his father is Irish, his mother Icelandic. He grew up partly in England and partly on a tiny island off the coast of Iceland famous for its puffins. For him, like many hackers, the Internet connection was his portal to the world.

In early January 2010, Smári flew back to Iceland with his list of laws and compiled them onto a wiki, an online document accessible to others. He and Birgitta began talks with parliamentarians while the word went out to the international info activist community that history was about to be made and did they want to come and help draft the legislation to create the world's first data haven. Julian arrived around 5 January; Jacob Appelbaum the hacker and Tor developer came next along with Daniel Domscheit-Berg; then finally the Dutch hacker, activist and entrepreneur Rop Gonggrijp, forty-two, who had known Assange for many years. They joined other Icelandic activists. The sleeping arrangements were haphazard for a while until they booked into a hotel suite in downtown Reykjavik. For the next two weeks the group met to thrash out their great idea.

They expanded the wiki and used a Web application, EtherPad, to work on documents simultaneously. The documents became rainbows as the various users made their own contributions. They would chat online even when they were in the same room. WikiLeaks brought their hands-on experience of using laws in other countries to keep material online and a growing list of media contacts; Birgitta knew about Parliament; Smári and the Digital Freedom Society activists knew about digital laws globally; and Swedish MEP Amelia Andersdotter helped to synchronise the Icelandic laws within an EU framework.

Originally Julian and Birgitta wanted to introduce a bill into Parliament but parliamentary officials advised against this, saying it would take more time and face more obstacles than if the idea was passed in principle through the mechanism of a resolution. Birgitta had plenty of obstacles already, not least the fact that she was an MP of a minority party, not even the opposition but a fringe party called the Movement. The likelihood of a non-governmental coalition member getting a bill through Parliament was almost zero. So the group redrafted their document into a resolution, which although it had no legal mandate, would show the will of Parliament. They wrote a series of core statements followed by a long annex which described in greater detail what changes in law were needed to create the Icelandic data haven.

The most tedious part was translation. They would come up with a draft in English, then translate it into Icelandic, after which it would be picked over at the Althingi by the parliamentary clerks, who would make suggestions or changes. These were taken back to the group and translated back to English; the document would be reworked, and the translations would begin again. This happened about eight times during two weeks.

Finally, when they had everything nailed down, they looked at the final document and it was a mess. An element of beauty had been lost in all the translations.

They decided to throw it out and start from scratch.

There was a collective groan from everyone, but they also wanted the legislation to be the best it could be. It was setting a new standard for the rest of the world so it ought to be perfect. They crowded back into the suite at the Foss Baron hotel, pungent now with an aroma of sweat and stale food, and began working into the night. The smell was getting a bit much for some. Daniel had lived with Julian for some time, but such cramped quarters highlighted his friend's capricious attitude to bathing. He also wore the same clothes, unwashed, for days at a time. Daniel had always accepted his colleague's many eccentricities because he believed in WikiLeaks and he could understand why Julian might think personal hygiene and clothes were trivial compared to the work they were doing. But when they all had to live and work so closely . . . well, that was different. It was not very considerate to the other people in the room. Possibly for the first time, Daniel took a close look at Julian. He was totally absorbed in his computer. His hair was greasy, his clothes askew. Daniel probably knew Julian better than anyone else and yet, looking at him now, he didn't feel he knew him at all. Julian could be charming, he could be brilliant, but what was in his soul? Daniel had no idea. He looked down at Julian's socks. They were the same pair he'd lent him six months ago, now with holes, and the soles imprinted with a dark footstep. He knew this was exactly how Julian would turn up to meet the politicians tomorrow.

'Julian, if you are going to meet politicians perhaps it is an idea to dress up a bit,' Daniel said.

Julian pretended he hadn't heard.

'I'm just saying. It's not an insult, it's just to be practical. When you dress like them it makes them feel more comfortable and that is our aim. We want to do everything we can to get this law passed.'

Julian looked up and gave Daniel an icy glare. Maybe he was ashamed, maybe his pride was hurt; but not so much that he changed his socks. Whatever it was, Julian decided at that moment, perhaps not even consciously, that he was finished with Daniel. He was not to be trusted. He was a hostile force. Daniel knew too much, not just about WikiLeaks, but more importantly about Julian himself. There were grave consequences for anyone who penetrated this far. Daniel didn't know that at the time but he would soon find out. The next morning they made the final translation back into Icelandic and the resolution for the Icelandic Modern Media Initiative is submitted to Parliament.

On 25 February Birgitta Jónsdóttir speaks for the bill and it passes. Then the real politics begin, in the hallways, in the committee rooms, where hopefully the resolution will return unscathed for a vote before the full Parliament in June 2010.

Daniel and the others leave, but Julian stays on, with Birgitta as his new confidante. He meets her kids, he even borrows her teenage son's clothes. They are spending a lot of time together, so it's no surprise that Birgitta notices when Julian becomes even more obsessed than usual with his laptop. He is intensely focused on something.

'What is it you're watching all the time?' Birgitta enquires one evening.

Julian thinks for a moment, then pushes his laptop across

so she can see it. He types in a password and a grainy black-and-white movie starts playing. But this isn't *Casablanca* or some other Hollywood classic. It's the raw footage taken from an Apache helicopter of American soldiers gunning down people on a Baghdad street.

'My God. I've never seen anything like that.'

'Neither have the American people,' he responds. 'That's about to change.'

4

Welcome to Wiki Wonderland

Tønsberg, Norway, Saturday 20 March

The usual method for writing a book like this is to interview experts, read everything on the subject, investigate, pontificate, synthesise and write one's findings in detached fashion. But a small Norwegian town on an icy harbour marked the place where I fell into the rabbit hole and entered my own story.

I'd been invited to speak at SKUP, an annual gathering for Norwegian journalists where 600 reporters and editors from Norway and Scandinavia and guests from Europe and the US gather for three days to meet, conspire, learn and teach. The focus this year would be the future of investigative reporting, a subject in which I had a direct interest. A friendly Norwegian newspaper editor by the name of Jan Gunnar Furuly wanted me to come and give a talk about my five-year freedom of information campaign to prise open the UK Parliament which had led earlier that year to one of the biggest scandals in British political life.[1]

1. The scandal resulted in the resignation of the Speaker and six ministers, with more than 120 MPs stepping down at the General Election in May 2010. An ongoing police investigation has so far led to convictions of three MPs and one Member of the House of Lords.

I was intrigued by the conference and when Jan mentioned that Julian Assange from WikiLeaks would be speaking too, I agreed immediately.

I'd been mulling over the idea of a digital revolution for some time. When the MPs' expenses scandal broke I saw first-hand that despite what many pundits said, people *were* interested in politics – provided they had meaningful information. Yet that was precisely the commodity in short supply – a situation mirrored around the world. We get plenty of propaganda but meaningful facts are few and far between. In Britain, information held by the powerful was not shared equally but rather divvied out through secretive networks of patronage and favouritism, and it was this information asymmetry that kept the political system in place. As I dug into the operations of the British Parliament, I found an institution set in aspic: a centralised hierarchy with a hostile attitude towards the public. Journalists either didn't care to look or accepted the Panglossian views of those in power that the way things were was the way they would always be so there was no point pressing for reform. WikiLeaks impressed me because they clearly believed change was not only possible but necessary.

I didn't know what to expect from Julian Assange. On Saturday I saw him for the first time. He sported a *Matrix*-at-home look, wearing brown utility trousers and a brown blazer, a black shirt and a thin, bright red tie. He's lean and tall at 6 foot 2 inches but the striking feature about him is, of course, his platinum-blond bob. WikiLeaks was still very much a niche interest at this time, reflected in the sparse attendance of Julian's morning session.

'Is this it? I'd expected more,' he says, looking around at the eight of us. If the small audience bothers him it doesn't affect the passion and vaulting idealism with which he speaks about WikiLeaks' battles to publish the unpublishable. He is unashamedly high-minded about free speech and the people's right to know. 'We intend to place a new star in the political firmament of man,' he announces.

He is forthright and uncompromising. 'The UK is the worst liberal democracy in the world,' he announces. 'It never went through a revolution so it's still a feudal state. The laws were made for the benefit of the lords and aristocracy. Now they are for the benefit of the new lords, the political elite.'

Asked how he came up with WikiLeaks, he says it was clear to him that the sheer amount of information generated in the world is no longer capable of being published by traditional journalists. There are too many laws hindering investigative journalism, and too much of what is hidden from the public is not about national security or confidentiality but protecting personal power and private interests. 'We exist to protect the public interest. We are a new spy network: the Public Intelligence Agency.'

A reporter asks how WikiLeaks deals with legal threats, which are a constant problem for investigative journalists. He says WikiLeaks has received about a hundred legal threats in the past two years, but that the technical structure and legal arbitrage of WikiLeaks means they are based everywhere and nowhere. They are beyond any one jurisdiction and thus cannot be stifled. 'We publish those lawyers' letters too,' he says, a mischievous smile breaking across his face. 'We look at them as just more free content.'

He wants to see journalism done the way science is done so that raw source material is available for verification. 'The strategy of media manipulation is to dump so much news on a cash-strapped media that it's no longer fiscally possible for them to generate their own content. Reporters rewrite what the government gives them. It's censorship but of a more nuanced style.' The duty of a journalist, he says, is to get primary source material into the historical record. It's also to ask tough questions and remain sceptical – but I'm impressed by what he says.

Listening to him speak, I feel elevated, a great expansion of what is possible. I grab him for an interview after we've both finished our presentations, and I do literally take him by the arm. It's not something I usually do but he seems elusive and ephemeral, liable to disappear at any minute.

'I'm trying to find a place where my back isn't exposed,' he tells me.

'That's tricky. You're basically in a glass house.' The main lounge has a bar on one side and a wall of windows on the other giving a panorama of the river, glittering with late-winter ice chunks.

'Let's go over there,' I say, and lead him over to the non-windowed part of the lobby.

That evening I catch up with him in the hotel lounge and he's seated on a sofa in front of the very same floor-to-ceiling windows. Outside, oil lamps line the riverside walkway, their flames licking the Norwegian night sky. A few well-wrapped Norwegians sit outside on metal chairs covered with sheepskin, their faces lit by candles encased in glass bowls. They look almost cosy but for the clouds of frost that appear whenever they exhale.

Julian is wearing the same outfit as earlier, so getting ready for tonight's gala dinner and awards ceremony must not be a priority. He's typing on a little laptop and there's a small silver briefcase on the low glass table. I slide into the chair next to him.

'So what happened to the sniper?' I venture. 'Do you think they got bored trying to take you out?'

He smiles shyly, looking up through his lashes, like a coquette.

'Want a beer?' I ask.

'Sure.'

I head over to the bar and look back. He smiles. I smile back. What a strange guy – but certainly the most interesting person I've met for a while.

When I'd mockingly asked that morning if he thought he was going to be shot, he'd been silent, which I took to mean it was a serious possibility. Why come to an investigative journalism conference if that's the case? It's a good question so I ask it plus a follow-up: why does he think there are snipers with their gunsights locked on him?

'I've been tailed from Iceland by two US State Department officials.'

If this is true (to me it seems highly dubious) then I wonder why he'd tell it to someone he's only just met. After all, *I* might be the tail. It's the sort of statement you might hear from a lunatic, a fantasist, a narcissist or a show-off – or possibly all four.

I ask him what evidence he has for his claim.

'I have friends working in airline check-in desks and they alerted me.'

'How do they know? I mean, it's not like spies announce themselves on flights.'

He says two people were on his flight with diplomatic passports. Their seats were held but they didn't give their passport information until the flight was boarding.

I'm starting to get the impression that to be with Assange is to enter a political conspiracy thriller, where the plot is so wildly convoluted and the action so fast-paced that logic is forgotten. Whether his paranoia is warranted or not, I can't yet tell. As someone who has spent her life covering police and politicians, I tend to be pretty sceptical. On the other hand, investigative journalists get a lot of people contacting them who seem crazy or obsessed but who often have amazing information.[2]

'Why do you think you're of such interest to the State Department that they'd have you under surveillance?' I challenge.

'I can't tell you.'

But later at lunch, he does. 'You should come to Washington DC on 5 April. We're having a press conference.'

'About what?'

He leans in and says he has some amazing secret footage. 'It is shocking stuff that shows collateral murder by a major Western government.' Later he says it is the United States.

Again I wonder – if it's so top secret, why is he telling me, someone he's just met?

I steer the conversation back to more solid ground. I want to locate some of the documents he referenced in his early-morning presentation. Someone asked if he had any advice on how WikiLeaks talked their sources into leaking. 'We have published loads of manuals from secret services on that matter. We use their advice on converting walk-ins to long-term sources and mining patronage networks and journalists. Read them!'

2. I've had a woman phone me up with the seemingly outlandish story that a funeral home had mixed up her dead father's body with that of another man. She phoned in tears saying the funeral home had tried to pass off this other body as her father at the wake, dressing him up in her dad's clothes right down to his false teeth. It all turned out to be true and led to a review of the way funeral homes were regulated in South Carolina.

I have tried to read them but the entire WikiLeaks site has been down for a months-long fundraising drive. 'How can we read them when they're not online?' I ask.

He tells me everything remains on a mirrored site and when I hand him my laptop he types in the address. He shows the operations manual for the US detention centre at Guantánamo and several other reports, though I can't see any specific mention of source cultivation. 'You have to read them all. It's in there. You just have to look.'

I'm keen to learn the logistics of setting up a website offshore to avoid being censored. I obtain information through legitimate channels using the law, but it's a difficult and often thankless task. I can't help noticing that WikiLeaks is exposing a good deal more corruption and injustice through illegitimate leaks. But it's still dangerous. There are very real threats of arrest, litigation, bankruptcy and violence.

'How is it that you're not afraid?' I ask. 'I'm constantly torn between doing things that shake up the system and then feeling anxious because of what I've done.'

'It's natural to be anxious but you must do what you fear,' he replies, moving closer. 'Fear exists largely in the imagination. That's what powerful people prey on. It simply isn't possible to police the world's citizens so what they rely on is fear. Once you realise that fear exists largely in your own mind, then you are on the way to liberation. Also I have this righteous indignation. It spurs me on. Makes me feel invincible.'

All the while as he says this he's cat-like and aloof – until he looks right at me. I don't think it's anything I've said because he's quite uninterested in what I (or, seemingly, anyone) has to say, but at various points his focus shifts from detached and distant to direct and intense. He cocks his head to the side, raises an eyebrow and looks intensely at me. The transformation is remarkable and startling. When he has his eyes on me – as he did just now when he was saying that fear exists largely in our own minds – I have the sense he's looking right into my soul. The teenage girl in me swoons madly, but the investigative journalist concludes that the detached/intense thing is a technique he's honed after years of practice to get people to open up and give away their secrets. I have to admit it's pretty effective.

'The premise of my book,' I tell him, 'is that we are in the midst of a digital revolution. The interactive Internet is providing the mechanism for the world's first truly global people's uprising and democratic revolution.'

He shakes his head. 'The Internet is the easiest place to censor.'

How can that be?

'At the moment there is freedom but all the time it is eroding.' He describes the three stages that a state goes through to transform itself from free to authoritarian.

First, an argument is made that there are 'bad things on the Internet' so something must be done. Second, politicians increase their standing and power by passing a censorious law so they appear to be dealing with these bad things. Finally, as a result of this law, an industry of censorship develops which then has a vested interest in generating more demand for censorship products.

He mentions Australia, Germany, the UK and even Norway as Western democracies in the process of, or already having, Internet filters which censor sites secretly. Consequently the majority of sites blocked are not actually dangerous. 'We published Australia's blacklist and there were all sorts of sites on there that had nothing to do with child pornography or anything illegal.'

Once he gets into his stride I reach for my notebook. I have the feeling that every sentence he speaks has been spoken first in his mind, and rolled around for style and succinctness until it pops out perfectly polished, like a mathematical formula. He makes much of his studies in maths and physics, though later I discover he only did a few courses at the University of Melbourne and actually has no qualifications of any kind.

He speaks a lot in slogans. On achieving substantial political change he says, 'All you have to do is not give up.' On dealing with opposition, 'Power always pushes back. If you're not getting pushback then you're not fighting hard enough.' It's as though he's already imagining how such aphorisms will one day be collected and used to explain the birth of the future age, a time when individuals and information roam free, unhindered by secrecy-loving despots. Indeed, sitting here like this, I am starting to feel like a disciple. I'm not prone to this sort of behaviour. I think what clinches it is his complete lack of self-doubt and his fearlessness in the face of power.

'Have you ever been arrested?' I ask.

'Of course. Several times. I was a bit worried when I was stopped by the secret police in Malaysia. They arrested me for what I'd said about Anwar Ibrahim [the opposition leader of Malaysia accused of sodomy]. I thought that was going to be pretty bad but it was no big deal. Easy really. Once you've done it once you lose your fear. It gets easier. Then before you know it,' he says with a child's smile, 'you'll be a megalomaniac like me.'

When I return that evening with our drinks, two more journalists have sat down. I notice, now that I'm close to him, that his long pale fingers have black crescent tips and his silken hair is somewhat matted. We talk about some of the panels we've attended but I'm keen to get to the bottom of

Assange. I feel compelled to know everything about him. I start with the most obvious: is that his real hair colour or is it bleached?

He begins to spin out some stories. 'Yes, it's real. I used to have it down my back. When I was travelling in Vietnam it got very annoying because everywhere I went people would come up to me and want to touch it. So I decided the next time I went travelling in Asia, I'd dye it black. So guess what? I went to Japan and did exactly this only to discover that in Japan people dye their hair all sorts of crazy colours and in the manga tradition the male hero usually has white hair.'

He does have a comic-book hero look. When I ask if it's always been white he says no. 'It went white as a result of a childhood experiment with a cathode ray tube that went wrong.'

I recall the blue superhero in *Watchmen* and wonder if he's populating his biography with comic-book references.

'And your name? Is it French?'

'Some people think it's French or African. My mother is French. My grandfather was a Taiwanese pirate.'

'Of course he was!' I'm laughing now.

'He really was.' He looks a little hurt that I might not believe him, which I don't. 'He was a pirate and landed on Thursday Island where he met and married a Thursday Islander woman. They went to Queensland.'

'Is there any part of your life that isn't mythical? Your story has enough plots for a dozen thrillers.'

'Yes, I read your tweets.'

All my sceptical instincts are on alert. The changes in story – one minute about to be hit by a sniper, the next sitting in front of those huge windows. The paranoia. The concoction about his hair colour, and that he's been sitting here alone reading tweets about himself. These things strike me as reasons to steer clear. Yet when he speaks it's a siren's song.

Later I get to the meaty stuff: 'What drives you?'

When Julian offers only silence to my question I probe further: 'Does your anti-authoritarianism come from rage against your father?'

He turns to me with raised eyebrows. I can't tell if he's offended or curious. But surprisingly he answers.

'No. The man I thought was my father for most of my life turned out not to be. I didn't find out who my real father was until I was twenty-eight.'

'Your mother didn't tell you?'

'No. But I got along with my stepfather. When I met my real father, though, it was obvious I was his son. We're exactly alike. It's a clear indication of the power of genetics.'

'But still – not knowing him until you were twenty-eight. Why didn't he ever get in touch? That must have made you angry?'

He shakes his head. 'We share the same rebellious nature. Both my parents – they were protestors in the Vietnam war. I can understand why he wasn't interested in looking after a kid.'

'And are you close now?'

'Yes. We see each other occasionally.'

'So if it's not about your father, where do you think your drive comes from?'

He thinks for a minute. 'I was looking at the email of Pentagon generals when I was seventeen. That does something to you when you're a young man. That's a defining experience. I think you can see what sort of a person you'll become based on your first experience with power as a young man. That was mine and once you've had power like that it's hard to give it up.'

It's an interesting theory. My first experience with power was as a university student looking through politicians' expenses. 'Maybe that's why I'm compelled to investigate politicians,' I say.

'No. It's different when you're a young man.'

'Why do you keep emphasising "young man"? Do you think women can't be driven in the same way?'

'No. They're not.'

It's said as a definitive statement although I notice he feels no obligation to provide any supporting evidence. I'm rather stunned he'd think this, let alone say it. I stare at him baffled and not a little disappointed. He sips his beer, oblivious.

Also at the conference is David Leigh, the investigations editor of the *Guardian* newspaper. He's in his sixties and one of the foremost investigative reporters in Britain. His reporting put former Conservative defence minister Jonathan Aitken behind bars for perjury, and exposed international cigarette smuggling by British-American Tobacco and bribery over arms sales by BAE Systems. He started his career in the 1970s and, as some women keep the hairstyle of their youth, so Leigh keeps not only the hairstyle but the glasses, leather jacket and turned-up collar of investigative journalism's heyday. His face may look like a fallen soufflé but his eyes are as beadily alive to stories as any cub reporter. We're at the same table at dinner and both marvel that a country as tiny as Norway can pull in 600 reporters and editors to talk about investigations. The conference organiser Jan Furuly says the turnout is nothing special.

'It springs from a Norwegian idea called *dugnad*,' he tells us. The idea is that everybody in the journalistic profession shares their experiences for the benefit of all to improve the quality of the journalism. 'In Nordic society we're raised to believe we are equal with our countrymen and -women. In the more Continental and Southern European sphere, not to forget the British, there are stronger demographic differences in classes and regions – and to my experience a more sceptical approach towards the idea of sharing anything with colleagues outside their own organisation.'

The Norwegians are so committed to sharing that they have travelled around the world to proselytise their common-wealth of journalism. At least one tangible result is that a whole new generation of journalists in Estonia were trained by Norwegians and Estonia's freedom of information act is based on the Norwegian law.

There are two interesting facts about Norway: it is the world's number one democracy (according to the *Economist*'s 2010 Democracy Index) and it has the highest number of newspaper readers per head of population.[3] I'm not certain if a causal relationship can be drawn here but anecdotally there does seem to be evidence that the greater the number of media voices, the stronger a democracy. There are over 200 paid-for newspapers and 15 free papers. All cities of Norway have their own local paper, which most households subscribe to; in addition most read at least one national or regional larger paper. When the rest of the world faces news closures, how is it that Norway can sustain this level of journalism?

Jan's answer is that Norway has a 150-year tradition of state

3. Although Norway has a market of just over 5 million people, it tops the World Association of Newspapers' list for the number of newspapers read by an average person. The online newspapers are equally popular, with the largest news site, VG Nett, having a daily reach of 37.6 per cent of the population in 2011, Dagbladet reaching 24.4 per cent and Aftenposten.no 14.8 per cent.

support for the media. Since the 1960s there has been a policy to support the weakest newspapers in each city. In 2010 the state paid out close to 270 million NOK (£29 million) to the least-funded papers. There is also indirect public support through a sales tax (VAT) exemption for newspapers, which particularly helps the biggest newspapers like *Aftenposten* and *VG* (*Verdens Gang*). The total amount saved from the exemption is some 1.5 billion NOK (£165 million). There is also a public broadcaster – the Norwegian Broadcasting Company, Norway's equivalent to the BBC, also funded through an annual fee of 2,478 NOK (£274) paid by every household with a television set. The amount generated (4.5 billion NOK, nearly £500 million) pays for a service of 3,500 staff.

In Britain and America such state sponsorship is frowned upon. It may not be the answer but the public is already heavily subsidising media in both these countries. The only difference is that instead of going to an independent media, these subsidies are going to government PR units to peddle state-sponsored spin. The UK's Central Office for Information had a turnover in 2009/10 of £532 million and has consistently been one of the biggest spenders on advertising (number one in 2009, dropping to fifth place in 2010). Local councils across the country publish their own newspapers which in many cases have undermined independent papers. While few US local governments could get away with running their own taxpayer-subsidised newspapers, subsidisation does exist: the US Army spent $4.7 billion and employed 27,000 people on public relations in 2009 according to an Associated Press investigation. If we must state-sponsor, at least let it go direct to media organisations for reporting, as in Norway, rather than government bureaucrats.

Journalism's USP

Hearing Assange's ideas about how the press needs to change, in the midst of such a concentrated population of journalists, made me wonder – what is the role of journalism in the digital age? Certainly there has been much hand-wringing within the profession about the 'death' of journalism in general and investigative journalism in particular. Let me here set forth a rallying cry for proper journalism, by which I mean the journalism of verification.

Professional journalism has one real Unique Selling Point in the digital age, one reason why anyone would want to get their news from a professional journalist[4] rather than anywhere else: the ability, gained through training and experience, of sifting through mounds of information for what is a) important and b) verifiable as true.

At a time of information overload, good journalists are more important then ever. They serve as the public's hired guns to collect information from various sources and challenge it for the purpose of distilling down what is important and true. They signpost issues that are worthy of our attention. In the past when we bought newspapers we were paying for that particular newspaper with its content – a bundle of news and entertainment. In the digital age we're buying the carriage (e.g. the Internet access) and readers decide later what information they want to view over that carrier. The choice is enormous. Between 1990 and 2005 more than 1 billion people moved into the middle class, according to the *Economist*, and as they did so they entered the information age, uploading and

4. Being a professional journalist is rather like training to be a lawyer. There's a certain amount you can learn in school but largely it is a vocation learned through practice, with scepticism being a primary attribute.

downloading information at an unprecedented rate. Already in 1970 the author Alvin Toffler was predicting in his bestseller *Future Shock* that mass acceleration of knowledge and technological innovation would overwhelm people.

> In three short decades between now and the 21st century, millions of ordinary, psychologically normal people will face an abrupt collision with the future. Citizens of the world's richest and most technologically advanced nations, many of them, will find it increasingly painful to keep up with the incessant demand for change that characterizes our time. For them, the future will have arrived too soon.

Toffler wrote about the uniqueness of living in the '800th lifetime', which he calculated based on the last 50,000 years of human existence divided into lifetimes of sixty-two years. For 650 of these lifetimes we were living in caves. Only in the last seventy has it been possible to communicate across generations through writing and only in the last six have people seen a printed word. Only during the last two has anyone used an electric motor and finally – only in the last 800th lifetime – have most of the material goods that we use in everyday life come into existence. The Internet and social networking are a fraction of this lifetime.

What is extraordinary about this isn't just the scale of change but also how intimately we all experience it as we grow ever more interconnected through digital networks. It used to be that a major disaster in Japan affected only the people of that country. Now we can download videos of a tsunami hitting Japan within minutes of it happening. We see the Egyptians protesting in Tahrir Square and taking over the headquarters of the secret police. We see citizen videos of Saudi troops entering Bahrain. In the midst of this there are rumours, fakes, lies, misunderstandings, obsolete data and changing opinions. Even writing this book I am acutely aware of the rapidity of change and how much things are in flux. To hold steady into the future, it's necessary to live as though floating above it all to gain some perspective on what is transient and what is less transient in order to determine what is important.

The powerful have historically tried to impose their will through mechanisms of enforced ignorance such as censorship, secrecy, threats, physical intimidation and violence. This model is difficult to sustain in a networked world based on Enlightenment values. This is not to say that Western democracies have abandoned these heavy-handed tactics, but more often the methods have shifted to more sophisticated ways of maintaining power such as media management, public relations and legal intimidation. In the midst of all this information and misinformation how can we filter out what is important and true?

When a politician claims for example that 'crime is down' since he implemented a certain policy, it is the professional investigative journalist who knows the raw data on which this statement is based (criminal incident reports) and who asks for verification. He or she can then go to other sources to question the veracity of the data. The reason I specialise in the intricate details of bureaucracy isn't because I have a passion for paper-pushers, but rather because I need to know all the types of information collected, by whom and where they are stored so I can get my hands on them. A statement isn't a fact. Even when the person making the statement is an authority he or she still needs to provide evidence or proof that what they say is the truth and a professional journalist should be asking for this proof and supplying it for public scrutiny.

All this accumulating of statements, data and information

which then has to be verified takes time. But this is the *only* thing a journalist does that marks him out as a professional. It's the only reason anyone would choose a well-known newspaper's website over an unknown blog. The newspaper as a brand has built up, over time, a reputation for challenging the powerful and giving people meaningful, true information.

The press is not like any other business and what it sells shouldn't just be rehashed press releases or celebrity gossip, but the civic information necessary for people to understand their society and participate in it. It is a check on political and financial power, or at least it should be. The people of Iceland saw clearly the consequences of a press no longer fulfilling its civic function.

Newspaper circulation everywhere is on the decline and viewing figures are fragmenting. But this doesn't indicate a public uninterested in meaningful information. That would be to conflate the medium with the message. The Internet has lowered the barriers to publication so competition is much more intense. Unfortunately, many media owners have responded to this competition not by focusing on their USP but by ditching it altogether. Reporting budgets have been slashed and papers are increasingly filled with reprinted press releases from the government and PR agencies – *churnalism*. As Nick Davies explains so well in his book *Flat Earth News*, the journalism of verification has been largely abandoned by the owners of the modern media conglomerates (who then wonder why no one wants to read these news-free newspapers).

When there is meaningful information that affects people's lives and they are in a position to exert real power based on this information, they are interested. When I was digging into Parliament, I was repeatedly told by editors that the public were not interested in politics. This proved to be untrue: when the expenses data finally came out the public couldn't get enough of this story. No MP was too minor for his or her constituents to take an interest. A perception of indifference doesn't equate to a lack of interest, it is often a symptom that the public either don't have meaningful information or that they are disenfranchised from doing anything with it.

Halfway through dinner that night we notice Julian has disappeared.

'Go on – you call him,' Jan Furuly says. It's odd how Julian has created desirability by making himself scarce. Even David Leigh has picked up on it. We all want to talk to the WikiLeaks man.

I get hold of him and we all reconvene later in Jan's suite over a 23 year old bottle of Zacapa rum from Guatemala. We talk about the hacked emails of the University of East Anglia which revealed that the university had breached the Freedom of Information Act when handling requests from climate change sceptics (the university escaped prosecution because the case came to light outside the six-month time limit for cases to be brought). Just as we're about to head to the end-of-conference party next door, Julian tilts his head sideways and says to David, 'I have something to show you.' From behind his glasses David's eyes glisten. He follows Julian down the hall.

Afterwards David tells me how Julian took out his little laptop in the hotel room, entered a password and showed him a video taken from the gunsights of a US Apache helicopter in Iraq, the same one he'd shown Birgitta a month earlier. David was stunned, so much so that he wasn't quite sure what to make of it, or of Julian, who disappeared again shortly after showing him the explosive footage. The *Guardian* had worked with WikiLeaks on two stories previously (Trafigura and

Barclays Bank's tax avoidance scheme) but this was the first time anyone from the newspaper had met anyone from WikiLeaks in person. David is quite blunt about how he deals with people. 'It's not about friendship. I'm interested in people for the information they possess. Julian struck me as someone with a lot of information and therefore he was very interesting to me.' But reliable? David wasn't sure.

The *Guardian* editor's suspicions were cemented an hour later, when Julian came banging on his hotel door.

'What have you done with it?' Assange was shouting.

'Done with what?'

'My computer!'

David hadn't even realised it, but Julian had left his computer in the editor's hotel room. Dishevelled and highly embarrassed, Julian grabbed it, cast his grey eyes around the room and fled once again into the night.

5

In the Belly of the Beast

National Press Club, Washington DC, Monday 5 April

It's nearly 2 a.m. when the Acela train pulls into Washington DC's Union Station. Julian Assange and Dutch hacker Rop Gonggrijp find a taxi and head along an eerily empty Massachusetts Avenue lined on either side with bland office blocks.

'Here we are in the lion's den,' says Rop, looking at a federalist brick and stone building to the right.

'Not looking too lionish,' says Julian.

It isn't just the time of night. The American capital doesn't have the same energy or raw power as New York, where they'd landed several hours earlier from Iceland. Julian says he's been inside many of these institutions' computer networks, but he's less interested in their physical reality. There is one place though . . . He thinks about directing the taxi driver across the river to the Pentagon but no, there isn't time. The taxi pulls up a few hundred yards from the hotel (Julian didn't want to reveal where they were staying), the two men sling on their backpacks and Rop pays the driver.

Several hours later, it's a sunny spring morning and they're back on the street, now crowded with suited men and women on their way to fill the office blocks. Rop and Julian arrive at their destination and look up past the American flag at the fourteen-storey gold-brick edifice. They feel they are about to make history and indeed the National Press Club's slogan is 'The Place Where News Happens'. That's what they're here to do. This particular Easter Monday is Year Zero for the people's right to know.

It's a holiday in many countries but Julian reasons that will increase coverage. 'It's a slow news day,' he's told Rop. They're led to their assigned conference room and Julian gets out his laptop while Rop checks the projector and sound. A few reporters take their seats. Julian has a speech ready based on some notes he's made that morning on a crumpled piece of paper. He wears the same outfit as he'd worn in Norway: brown jacket, black shirt, thin red tie.

When Julian set up WikiLeaks back in December 2006 he'd thought it would only be a matter of months before it gained worldwide prominence. Instead, everything he'd thought would be easy was hard and the things he'd thought would be hard were easy. Getting around the world's various laws that suppressed information had been pretty straight-forward. He'd adopted the model of a tax-avoiding multi-national corporation, passing information through countries to take advantage of their laws, but without registering anywhere as an entity so he had the added advantage of telling any lawyers that came calling to go to hell. Getting people to notice WikiLeaks, however, was proving very diffi-cult. In the past, he'd thought the importance of the material would draw attention. It hadn't. Instead, documents languished online and only came to the public's attention when they were written up by professional journalists. The leaked US Army's intelligence analysis of the April 2004 Fallujah attack posted on WikiLeaks, for example, was ignored until the *New York Times* and the *Guardian* wrote about it. Raw material alone wasn't enough. People needed signposts supplied by the media. To get attention, Julian decided he must court the press.

'What you're about to see is the raw brutality of war,' he tells the thirty or so gathered reporters. It's a smaller turnout than expected but he's confident the news will spread once they see the footage. 'This classified US military video shows the indiscriminate slaying of over a dozen people in the Iraqi suburb of New Baghdad, including two Reuters news staff.'

Attempts by Reuters to obtain the video through the Freedom of Information Act had been unsuccessful. The Pentagon's account of the 12 July 2007 air strike is remarkably different from the reality they are about to witness, Julian promises.

The lights are dimmed and Julian and Rop wait for the video to begin, wondering how the reporters will react to the project they've spent the past weeks putting together in a little house in Iceland.

If the reactions from those who had seen it already were anything to go by, the video would make a huge impact. Birgitta Jónsdóttir was profoundly shocked when Julian first showed it to her, as was David Leigh in Norway.

By March, Birgitta was extremely busy not just with the Icelandic Modern Media Initiative but also with the parliamentary committee investigating the collapse of Icesave, one of the failed Icelandic banks. Yet when she saw the footage she decided to devote her entire Easter holiday to co-producing the video. 'I wanted to give voice to the voiceless. To the war that was brought to their doorstep,' she told me later. In horror and then sadness she'd watched as an American ground soldier (later identified as US Army infantryman Ethan McCord) carried an injured five-year-old girl from the wreckage of a van sprayed with machine-gun fire.

There was a lot to do to make the film ready for publication: go through the footage frame by frame, piece together the detail of what it depicted and the units involved, work out who was who, translate all the military jargon, and figure out whether the Iraqis on the ground were armed or not. Subtitles were needed and had to be translated into various languages, a stand-alone website built and a place found where the film could be posted to withstand both political pressure and the amount of traffic Julian predicted it would get. In order to ensure 'maximum impact' he wanted a shorter, punchier 'edited' version of the film and a media campaign. Rop and Birgitta took charge, creating workflow charts to organise a handful of volunteers who'd come to Iceland to help.

Julian had a strong belief that he was going to change the world and that it was only a matter of time before the world came to recognise this. The leaks he'd received a few months before convinced him that 2010 was the year this would happen. But he was also vaguely aware that the grand ideas that existed in his mind often failed to live up to his expectations when he tried to implement them. He was all about the big idea. Mundane details such as airline tickets, bills, food, lodging, organising and even programming were for others. Hence, the actual production of the video was done by a team of about ten or twelve idealistic, driven volunteers, and WikiLeaks' first major press conference was organised by Rop; meanwhile Julian spent a large part of the crucial weeks before the launch talking about his big ideas past and present with *New Yorker* reporter Raffi Khatchadourian. Daniel was back in

Germany. Julian's newly favoured courtiers were Rop and Birgitta.

Rop came out 'to make things sane again', he said, and was happy to loan 10,000 euro to the project with no guarantee that he'd get it back. He'd known Julian for many years from the hacker scene and just like the Icelandic Modern Media Initiative he believed 'Project B', as their work on the video was called, was necessary to redress the balance of power between state and citizen.

Julian's motivations were harder to pin down. He hoped Project B would make people sceptical about official narratives given out by the authorities, particularly the US military. He told Khatchadourian, 'This video shows what modern warfare has become, and I think, after seeing it, whenever people hear about a certain number of casualties that resulted during fighting with close air support, they will understand what is going on.' But Julian's focus on his media biographer was so intense there was a danger it would hinder the film's completion. Birgitta wasn't too happy and asked the reporter a number of times to stop providing Julian with a receptive audience. Like the other volunteers, she was spending eighteen-plus hours a day in the little house at 33 Grettisgata in Reykjavik to finish the video for Julian's 3 April deadline.

Icelandic journalist Kristinn Hrafnsson and cameraman Ingi Ragnar Ingason were also part of the group. They knew of WikiLeaks from the Kaupthing Bank loan-book story that was broadcast on RUV, the state television channel where Kristinn worked. Kristinn had stayed in touch with Julian and was one of the privileged few to be shown the video. 'I realised this was serious and important information,' he said later. 'Nobody believes people on the ground when they say war crimes are occurring.' That's why footage of the type he'd just seen was so revelatory and he was willing to give up considerable amounts of his time to make sure it was made public. Kristinn and Ingi knew the video would have more impact if it could be verified, ideally by tracking down and interviewing the surviving children in Iraq and finding relatives. The lack of time meant it was almost impossible for them to get the required visa, but it came through at the last minute.

'We got the visa!' they said, rushing into the house. They went to a cashpoint and pulled out as much money as they could,[1] then flew to Iraq with the secret video. By another stroke of luck, their fixer located the two children injured in the air strike so they quickly did an interview with them and their mother. The video shows their father driving a minivan through the rubble of the Iraqi city and stopping to rescue Saeed Chmagh, forty, a Reuters driver and assistant, who is crawling injured along the pavement. The father gets out and together with another man they carry Saeed into the van. During this time the gunners in the Apache helicopter flying overhead have their gunsights locked onto the men and request permission to shoot. Even before the rescuers arrive one of the gunners is in a hurry to kill the unarmed man: 'Come on, buddy. All you gotta do is pick up a weapon.'

The Apache receives permission and unleashes a hail of 30-millimetre cannon, perforating the van as if it is made of paper. They give the location to the ground crew: 'Should have a van in the middle of the road with about twelve to fifteen bodies.' The shooter interrupts with the proud exclamation, 'Oh yeah, look at that. Right through the windshield!'

1. The reporters paid for the trip themselves though Julian said WikiLeaks would reimburse them. Kristinn and Ingi weren't reimbursed until some months later when they contacted Daniel Domscheit-Berg, Assange's former right-hand man, who arranged for them to be paid out of funds from the Wau Holland Foundation.

When the ground soldiers arrive they report two children in the minivan.

'Well, it's their fault for bringing their kids into a battle,' says one of the gunners.

'That's right,' his colleague replies.

The volunteers in Iceland crowded around the computer to watch Kristinn and Ingi's filmed interview from Iraq. The mother explains that the children's forty-three-year-old father, Saleh, was driving them to class when he saw the wounded man moving in the street and drove over to help him, only to become a victim himself of the Apache guns.

'That's why they're bringing their children into a war zone,' said Birgitta, furious, 'because it's on their doorstep.' Later she reflected, 'These people – the people in Iraq – have been trying to tell what happened and they have been ignored, treated as if they were living in a fantasy world. Once you see this video, then you can show others that this is what is happening in Iraq. This is why US soldiers are not winning the hearts and minds of the Iraqi people. The military establishment is outright lying about what is happening. How does that help anyone?'

Government information: for whose eyes only?

The US military believes it owns the Apache helicopter video and as such only it can decide who is permitted to see the footage. What the leaker did was not just immoral but criminal and endangered national security. Interestingly, one thing the US government can't do, which governments of many other countries can, is claim that WikiLeaks violated its copyright by publishing the material. This brings us to the issue of information ownership, one of the most crucial of the digital age.

While military operations require some degree of secrecy

for effectiveness, if there is too much, and too little public accountability for the use of lethal force, an environment develops that is ripe for the commission of war crimes. The balance must be right.

If the American people saw the video perhaps they would halt the injustices that were undermining American efforts to rebuild Iraq. After all, it was their money paying for the war and supposedly their decision to be in it. What could be more important than information related to a country's conduct in an occupied country? Ethan McCord stated in *Wired* magazine after the publication of the video, 'We've been there [Iraq] for so long now and it seems like nothing is being accomplished whatsoever, except for we're making more people hate us.' Don't the American people have a right to know *why* they are hated – if indeed they are? If they are hated because local people believe American soldiers are killing citizens indiscriminately, then that problem needs to be addressed. It cannot be addressed when it is kept secret.

Government copyright is about controlling the use and dissemination of information. But even the Founding Fathers accepted there were limits, and where the release of information could cause harm, different rules applied. It is in the military, law enforcement and intelligence service where this is most relevant. More specifically, the twin criteria for keeping something secret is that publication could lead to loss of life or could harm ongoing operations. In the case of the 'Collateral Murder' video, the military classified it as secret, meaning its disclosure would harm the nation. The classification system is based on the concept of 'national security' and the basic levels of classification in the US government – confidential, secret and top secret – relate to the perceived damage to national security if the information were disclosed.

The inherent problem is that those doing the classification tend to have a vested interest in secrecy as it allows their decisions and actions to be free from public scrutiny. We are in a frustrating scenario, for how can we know, without seeing the information for ourselves, if it is being withheld for legitimate reasons of harm or for other, most likely political, reasons?

The obvious answer is simply to say that we need to trust the government. But blind trust is what the founders of America warned against as it led to the dangerous concentration and abuse of power. America is unusual for being one of the few countries in the world to grant sovereignty for US government information to the people rather than to a monarch or bureaucracy. This is due to the Enlightenment ideals held by the framers of the US Constitution, particularly James Madison and Thomas Jefferson, who were both influenced by the 'Father of Liberalism' John Locke and who took the view that monopolies of any kind, whether for government information or business, were bad for society. 'In Britain there was a reaction against the monopolies handed out by the Crown on everything from playing cards to sweet wine,' says Professor James Boyle of Duke Law School, who has studied copyright regimes in the US and Europe. The founders saw no public benefit in these monopolies and believed instead that ideas should be free to spread. Thomas Jefferson stated this in a famous letter which has become a rallying cry for those seeking copyright reform:

He who receives an idea from me, receives instruction himself without lessening mine; as he who lights his taper at mine, receives light without darkening me.

That ideas should freely spread from one to another over the globe, for the moral and mutual instruction of man, and improvement of his condition, seems to have been peculiarly and benevolently designed by nature, when she made them, like fire, expansible over all space, without lessening their density in any point, and like the air in which we breathe, move, and have our physical being, incapable of confinement or exclusive appropriation.[2]

These two principles – preventing monopolies and encouraging the sharing of ideas – found their expression in the First Amendment. The 1976 Copyright Act *17 U.S.C.* § 105 places in the public domain all work of the United States government. Judicial opinions, administrative rulings, laws, public ordinances and other official documents cannot be copyrighted.

Britain, by contrast, operates under a system of Crown copyright whereby official information is owned by the monarch or the bureaucracy in which it is held. It certainly doesn't belong to the taxpayers. Most of the Commonwealth countries – Canada, Australia, New Zealand, Kenya and Fiji – still use this proprietorial system. The people must ask those in power for permission to use it and very often pay extensive fees for the privilege. The Queen was able to claim £200,000 in damages from a British newspaper that printed her 1993 Christmas message two days before it was broadcast, citing not breach of privacy or security, but copyright infringement. This was when the media was full of stories about the Royal Family's marital problems and it was a convenient way to muzzle the press.

Coming from England, some of the Founding Fathers had likely seen first-hand how Crown copyright was used by the ruling elite to limit access to and public use of government information. Licence agreements, royalties and restrictions on re-disclosure further limited the people's right to know what

2. Thomas Jefferson to Isaac McPherson, 13 August 1813. For full text, see http://tinyurl.com/22udzn/

those in power were up to. Ditching Crown copyright was revolutionary and the First Amendment's prohibition against abridging the freedom of speech or of the press was in large measure an attempt to protect the free discussion of governmental affairs from controlling interference. Thomas Jefferson and others wrote of their concern that if politicians and bureaucrats were given control over government information (which is the purpose of copyright) they would likely use it to advance political goals or stifle public discussion on political issues.

The idea that a government shouldn't be allowed to copyright official information didn't strike me as important until I moved from America to the UK and saw how Crown copyright continues to be used by those in power to control and restrict civic information in exactly the same way in 2006 as in 1776. Even now, most public bodies plaster lengthy copyright notices on their responses to freedom of information requests, continuing a long tradition of forcing citizens to ask the permission of government before they can share government information. When a government can copyright a publication, it can use this as a means of suppressing information that is embarrassing, inconvenient or challenging to official doctrine. In 2004, Parliament's Clerk of the Scrolls tried to shut down – for breach of copyright – a civic website (theyworkforyou.com) that allowed people to easily discover how their Member of Parliament voted. MPs didn't like this site at all – they found it embarrassing and inconvenient. The official record of Parliament, Hansard, is owned not by the people of Britain but by Parliament itself, so the clerk was able to make his threat using copyright law. He probably hoped this was enough to intimidate the developers of the website into toeing the official line. Instead, one developer, Tom Loosemore, looked forward to a high-profile case where the elected officials of a democratic country were suing to shut down a website for daring to tell the public how public officials voted. Parliament eventually backed down and agreed to give the website a licence. It was a victory of sorts, yet although the licensing system has been expanded considerably since then, the fact that it still exists at all means citizens have to ask their government for permission to use public data.

The idea of prohibiting government copyright is beneficial not only to political freedom but also to the economy. The ability to use and reuse official government data without seeking permission is a factor behind the remarkable growth of the US knowledge economy. The multibillion-dollar satellite navigation industry, for example, originated from free GPS data obtained easily from the US government. Meanwhile, Britain, the Commonwealth and Europe have laboured under a system of proprietorial copyright which has stifled innovation by requiring citizens not only to ask permission but also to pay vast fees for using information already funded by public taxes. In Canada, publishers complain that the burdens of dealing with the government bureaucracy surrounding Crown copyright and licensing has made reproduction of Canadian government information in new formats 'commercially unattractive'.

'The claim is that making activities recoup their own costs will allow us to expand the information that is made available,' Professor Boyle tells me. 'But that argument fails empirically. There is considerably more geographic and weather information available in the US, which gives it away free,[3] than in the

3. A similar argument is made about cultural information – art, music, movies, literature. The cultural world has long been embroiled in its own information wars, between those who believe content should be freely shared, and those who seek to control it. Leading the charge against the current system is US law professor and technology writer Lawrence Lessig, who founded Creative Commons in order to give artists more control

countries that use Crown copyright. European attitudes towards private commercialisation actually work against the idea of openness. In the US if the government hands out weather data for free and people make a ton of money off the back of it, everyone says "Great!", it's good for the economy, good for us, good for the company. It's a win-win situation. In Britain that's not the attitude. There's a sense that the company has got something for free and now they're making money off it. "How terrible! They're free-riding!" They don't see the overall economic benefit that comes from sharing information.'

While the US has managed largely to avoid these pitfalls, the digitisation of data is bringing new challenges as government officials realise the value of information in electronic format. Suddenly citizens have a powerful way to hold government to account. In both the US and the UK, the central government is the largest producer, collector, consumer and disseminator of information. This is valuable not just economically but politically, with major consequences on citizens' lives. Fierce debates are raging about the sort of data collected by governments, who has access to it and ultimately who owns it, for it is the 'owner' who has the final say on policies about access and reuse. Most of the information we need to be an informed electorate is housed inside government bureaucracies and these have a historical tendency towards secrecy. This instinct will only increase as the values and uses of government information expand. In the digital age we need a sound policy on government information to ensure that the overall and long-term

---

over their work, and whose 2004 *Free Culture* has become the seminal book on information ownership in the Internet age.

public interest is not sacrificed for the short-term convenience of a few officials.

Let's go back to the National Press Club and watch the video play out.

It opens with the blips and static from a walkie-talkie. There is a quote from George Orwell: *Political language is designed to make lies sound truthful and murder respectable, and to give an appearance of solidity to pure wind.* We see a black-and-white landscape of block buildings and palm trees overlaid with the white cross hairs of the Apache's gunsights. An ongoing conversation is heard between the gunners of two helicopters and ground control. The camera zooms onto a group of a dozen men milling around on a street. Two have cameras which the gunners mistake as weapons, AK-47s and a rocket-propelled grenade (RPG). To the viewer, in hindsight, the cameras look like cameras and the relaxed attitude of the men makes it difficult to think they are fighters, but in the heat of battle it's understandable how they might look to the pilots as though they are armed. The Apache opens fire and for twenty-five seconds we hear the staccato explosion of the guns. The distant men continue ambling around unhurt for several seconds until a wave of bullets cascades into them. They run in shock but in a few seconds they lie dead on the street. Getting hit by a 30-millimetre round isn't like taking a bullet from a handgun. The bodies are destroyed.

'Oh yeah, look at those dead bastards,' says one of the shooters.

'Nice,' says the other.

The soldiers spot a wounded man crawling on the kerb. This is Saeed Chmagh, the Reuters assistant. They see he's unarmed. That's when the van drives up.

The soldier in the Apache misreports the rescue, saying, 'We have individuals going to the scene, looks like possibly, uh, picking up bodies and weapons.' There are no weapons in sight. The gunners lock onto the van and after permission is given to engage they obliterate it with cannon fire. The ground crew arrive and we see Ethan McCord's rescue of ten-year-old Sajad Mutashar and his five-year-old sister Doaha.

After this there is a third part where the gunners report at least six armed men entering a partially constructed building in the city. They ask for permission to engage. 'We can put a missile in it,' the gunner says. It's unknown how many people, armed or not, are in the building. It's half-built but at least two unarmed people enter after permission to engage is given. The soldiers note them but proceed to drop three Hellfire missiles into the building. It crumbles into rubble.

Some reporters gasp at different points while watching the video – at the callous talk of the gunners, at witnessing the moment of death, at the rescue of the children and the gunner's response. It's certainly not the sort of thing they're used to seeing at the National Press Club. It hits hard. The obvious reason why the video was withheld from public view for three years is that the Department of Defense found it embarrassing, but embarrassment is not a legitimate reason to classify information. The reporters go back to their offices and call the Pentagon, asking for an explanation for the behaviour of the soldiers and the handling of the Iraq war.

Defense Secretary Robert Gates responds irritably, 'These people [WikiLeaks] can put anything out they want and are never held accountable for it.' He claims the video is like looking at war 'through a soda straw'. 'There is no before and there is no after.'

Assange sparks back, 'Well, at least there is now a middle, which is a vast improvement.'

Pressure builds and two days later, on 7 April, US Central Command publishes a series of redacted records on the case, including an investigation by the air cavalry and infantry units involved that states the air crew 'accurately assessed that the criteria to find and terminate the threat to friendly forces were met in accordance with the law of armed conflict and rules of engagement'. The report concludes that the attack helicopters acted within the rules of engagement and that the Reuters journalists were indirectly to blame because they were in the company of 'armed insurgents' and 'made no effort to visibly display their status as press'.

The argument begins to shift from the actions of the army to those of WikiLeaks. Perhaps Assange doesn't realise it but his insistence on polemicising the video, rather than letting the facts speak for themselves, has laid the groundwork for it to be discredited. Why was it called 'Collateral Murder'? That is Assange's judgement and doubtless murder was the opinion of many others after watching the footage, but it is possible to view the video from the soldiers' point of view and judge differently. Several of the volunteers in Iceland, including Birgitta Jónsdóttir, tried to talk Julian out of using such a provocative title, realising this obvious politicising would detract from the content of the footage. Most preferred 'Permission to Engage', but Julian had a political point he wanted to get across with this title: 'We want to knock out this "collateral damage" euphemism, and so when anyone uses it they will think "collateral murder".'

Assange is disappointed that the American media isn't more challenging of the Pentagon's statement that the killings were lawful. Fortunately, in the digital age he's no longer reliant on traditional media gatekeepers to get his message to the public. He posts the video (both the edited and full version) on a stand-alone website and YouTube. Soon it has more than 7 million views and WikiLeaks receives over $200,000 in donations.

But the Defense Department isn't going to concede easily. It launches a leak inquiry into this dangerous breach of 'national security'. The CIA and FBI announce they are very interested in questioning Julian Assange. What they can't do, however, is take back the video which is now circulating around the world.

6

Land of the Free?

Cambridge, Massachusetts, June

Danny Clark is starting to worry. He's heard nothing from his friend for more than a week: no emails, no messages, not even the usual quirky Facebook updates. Maybe Bradley Manning has been killed or injured? He isn't on the front line but he is in a war zone. Or perhaps he's just overwhelmed – he's been going through a tough time. He's been demoted and is about to be discharged from the army. His relationship with Danny's other friend, Tyler Watkins, has fallen apart after several years. Danny worries that Brad isn't taking it well.

Out of the blue he has a telephone call from a hacker in San Francisco, Adrian Lamo. Has Danny heard from Brad? Lamo asks. Danny says he hasn't. 'I'm also worried,' Lamo admits, 'so tell me if you hear anything.'

To try to take his mind off it, Danny goes to a party with some fellow students from the Massachusetts Institute of Technology. While there, he gets another call from Lamo.

'Manning's been arrested,' Lamo tells him.

'What for?'

Lamo says he's trying to confirm the story, but it seems to be about Manning being the source for a leaked army video that was published by WikiLeaks back in April. Lamo says he's doing publicity for Brad: 'I'm trying to put a positive spin on this.' *Wired* magazine are writing an article – would Danny talk to the reporter Kevin Poulson to give him some background about their friend? 'You can tell him that Brad's a good guy and this is surprising,' Lamo suggests. Danny agrees and speaks to the hacker-turned-journalist in a back bedroom at the house party. He gives Poulson various personal anecdotes about what a decent guy Brad is and about his family situation. 'But he didn't use any of that,' Danny said later. 'Obviously what he was trawling for was something along the lines of the quote attributed to Tyler – that Brad was wrestling with his conscience over the "Collateral Murder" video.'

'Did he confide to you that he'd leaked the "Collateral Murder" video?' Poulson asks Danny.

'No, he didn't. I don't know if he did these leaks.'

On 6 June the *Wired* article comes out: *US Intelligence Analyst Arrested in WikiLeaks Video Probe.* In it Poulson writes:

In January, while on leave in the United States, Manning visited a close friend in Boston and confessed he'd gotten his hands on unspecified sensitive information, and was weighing leaking it, according to the friend. 'He wanted to do the right thing,' says 20-year-old Tyler Watkins. 'That was something I think he was struggling with.'

Watkins reports that Manning had been back in touch after Julian Assange's Washington Press Club event:

'He would message me, Are people talking about it? . . . Are the media saying anything? . . . That was one of his major concerns, that once he had done this, was it really going to make a difference? . . . He didn't want to do this just to cause a stir . . . He wanted people held accountable and wanted to see this didn't happen again.' Watkins doesn't know what else Manning might have sent to WikiLeaks.

Over the next few days *Wired* and the *Washington Post* publish a series of alleged chat logs between Adrian Lamo and Bradley Manning that occurred from 21 to 25 May in which Manning 'confesses' to leaking not just the video but also US Army reports on Iraq and Afghanistan, personal files of Guantánamo detainees and 260,000 State Department cables. On 25 May, during a meeting with the FBI and army CID officers near his California home, Lamo turned Manning in, handing over the chat logs and his own hard drives.

The next day, Manning is arrested at Contingency Operating Station in Iraq. His computer hard drives are sent to Washington for forensic examination and he is flown to Camp Arifjan in Kuwait to await charges.

Manning first contacted WikiLeaks in November 2009 according to the chat logs. This was also the month when President Obama set forth his position on free speech during a visit to Shanghai's Museum of Science and Technology. Most of the questions he received from an audience of 400 Chinese university students hand-picked by Chinese officials were light and polite, but one came via the US embassy website asking, *In a country with 350 million Internet users and 60 million bloggers, do you know of the firewall?*

Of course Obama knew about the Great Chinese Firewall

that blocks and controls citizens' access to the Internet. He responded that he was 'a big believer in technology [and] a big believer in openness when it comes to the flow of information. I think that the more freely information flows, the stronger the society becomes, because then citizens of countries around the world can hold their own governments accountable. They can begin to think for themselves.'

His answer appeared almost instantly as the top news story on the official New China News Agency, Xinhua, as well as several popular Chinese websites. But anyone looking for it an hour later was out of luck. It had disappeared from the Net as though it never existed, as though it were dropped into the Memory Hole of George Orwell's *1984*.

In January 2010, Secretary of State Hillary Clinton expanded upon Obama's remarks with an entire speech devoted to Internet freedom.

'On their own, new technologies do not take sides in the struggle for freedom and progress, but the United States does. We stand for a single Internet where all of humanity has equal access to knowledge and ideas.'

She urged private companies to stand up to foreign governments demanding controls on the free flow of information or digital technology. She promised that the US government would support designers of technology to circumvent blocks or firewalls. She described a new kind of 'information curtain' that was descending across much of the world where countries erected electronic barriers to stop their citizens accessing portions of the world's networks. 'They've expunged words, names and phrases from search engine results. They have violated the privacy of citizens who engage in non-violent political speech. These actions contravene the Universal Declaration on Human Rights, which tells us that all people have the right "to seek, receive and impart information and ideas through any media and regardless of frontiers".

Rhetoric is not the same as action. The lofty Enlightenment values of America's founders have often been pushed aside when fear comes calling.

The Cypherpunks and Total Information Awareness

Encryption is used by human rights groups around the world to keep their communications safe from the prying eyes of authoritarian governments and secret police. It's exactly what pro-democracy activists in Iran and China need in order to ensure that the only people who read their messages are the intended recipients. Encryption allows a sender to encode a document or communication so only the intended recipient can decode it, using an authenticating 'key', and read the contents. This is essential because unlike postal mail where we expect a letter placed inside an envelope to remain confidential between sender and recipient (it's unusual for postmen to steam open envelopes these days), online documents and communications do not have the same confidentiality. Even in the United States they can be easily spied upon. PGP, which stands for Pretty Good Privacy, was the first encryption program available to the general public. It meant that while online communication might be intercepted by anyone, only the recipient could read the document. This was useful not just for activists but also for investigators, banks and anyone wanting secure communication online.

PGP was created by Phil Zimmermann, a military policy analyst, anti-nuclear campaigner and computer programmer who believed people needed the ability to protect their communications against hostile governments. He began working on PGP in 1984 and released his first version in June 1991. For version 2.0 he worked with other volunteers from around the world to make the program compatible with different platforms and languages. He had numerous phone calls and emails discussing source code, engineering and debugging issues with volunteers. All this was done in Zimmermann's spare time. Afterwards, his plan was to go back to policy consulting to try and clear up his five missed mortgage payments. PGP proved incredibly popular, but far from being praised for his efforts, Zimmermann became the focus of a three-year criminal investigation culminating in what became known as the 'crypto wars' of the 1990s. This was the first real battle of the digital information war and it centred on ideas about national security.

In 1991 a senate bill was also going through Congress that would have required all companies developing communications technologies to include back doors in their products for government interception. The bill was defeated but Zimmermann saw this as a confirmation that a Big Brother, all-seeing, all-listening government was not far away. People would not have any real privacy online against officialdom. It was a battle of power – the citizen versus the state – and it attracted a group of programmers and information activists who became know as the cypherpunks. Zimmermann's solution to the totalitarian future he feared was to write his own encryption program and give it away as free software. PGP spread quickly among peace activists and then made its way around the Internet.

In February 1993 Zimmermann received a telephone call from Special Agent Robin Sterzer of US Customs in San José, California, who asked him to explain what PGP was and how it worked. Zimmermann assumed they'd seized a computer loaded with PGP but couldn't decode it, so he was happy to help. But then the agent said she'd like to come and talk to him in person. He was living in Boulder, Colorado, at the time, over 800 miles away. It dawned on him that they didn't want him to help, rather *he* was the subject of a criminal investigation. He hired a criminal lawyer named Philip Dubois. The first time he went into Dubois' office he saw a box marked with the name of a notorious murderer and he wondered, 'What am I doing here? I'm in the office of a guy who defends murderers!' But Dubois turned out to be a good choice. He'd defended a lot of hard cases so was tough enough to take on the US government. Zimmermann set up a legal defence fund and other attorneys volunteered to join the case. The Electronic Frontier Foundation and the American Civil Liberties Union got involved. They all met to discuss Zimmermann's chances. 'They said I was going to prison,' Zimmermann recalls. 'That was the worst day of the three-year investigation. To have ten lawyers say it was hopeless.'

The problem was that communication of cryptographic software across borders was illegal. At that time the US government classed cryptographic software as munitions in the same category as stinger missiles, helicopter gunships or thermonuclear weapons, and its distribution was tightly controlled by export regulations. Even the discussions Zimmermann had with volunteers around the world were a violation of the Arms Export Control Act. There were exceptions to these rules – for example, you could export weapons about to be launched at an enemy – but unless they were going to put cryptography into a missile capsule and launch it into North Korea, PGP wasn't covered.

He was clearly guilty. But Zimmermann couldn't believe

that allowing someone to make their communications secure was a crime. He believed it was morally wrong that the state alone should have a monopoly on secure communication. Fortunately Dubois agreed, and as he was the only criminal lawyer in the room and this was a criminal case, his opinion carried the most weight. Dubois knew there are many ways to win a case, not all in the courtroom. If Zimmermann could win in the court of public opinion and show that the law was wrong, then it could be changed and he would be left alone.

There was another problem, however, this time related to the law surrounding information ownership. PGP used a public key algorithm for which a private company, RSA, had been granted a patent in the US. This effectively gave that company a monopoly on public key encryption, at least in America, and export controls stopped Zimmermann from developing a similar program abroad. The RSA version was based on a for-profit business model: every public key had to be certified by the company and paid for by the user. PGP, by contrast, was given away free and allowed anyone to sign anyone else's key. It was a company trust model versus a grass-roots trust model, and PGP's proved far more popular.

Zimmermann claims that Jim Bidzos, the president of RSA, was behind the criminal investigation, having contacted the US Customs Office in San José to request that they launch a prosecution. Several years later, Zimmermann met an official from RSA and asked why they hadn't just sued him. 'They told me it was bad PR to sue a folk hero. Better to get the Feds to put me in prison.'

Zimmermann was in trouble. On a trip back from Europe he was stopped at Dulles Airport by US Customs officers who took him into a white, featureless interrogation room. 'It was like one of those UFO abduction rooms,' he said. The agents had a dossier on him and asked questions: What was the purpose of your trip abroad? Did you do anything crypto related? Zimmermann refused to answer without his lawyer being present. He wasn't getting his lawyer, the agent claimed, and they would wait however long it took until he cooperated. They claimed he had no rights as he was at a point of entry,[1] but bravely he refused to talk, and waited. Finally they let him out and he called his lawyers.

While Zimmermann was being investigated, a curious thing happened. Another cryptographer, Phil Karn, wanted to export a book on cryptography which included the actual code of a program. He wrote to the State Department asking for a licence to bypass the Arms Export Control Act. The State Department said sure, it's a book, of course you can export it. It's protected by the First Amendment. This gave Karn an idea. A short time later, he sent in another request asking for a licence to export the exact same information, but this time on a floppy disk. The State Department realised they were in a bind. They'd already agreed to the book; how could they claim the same information was banned simply because it was digitised? They asked the National Security Agency what they should do, and the NSA told them to refuse. They did, and Phil Karn subsequently brought a legal case appealing the restriction as it was unconstitutional.

This, in turn, gave Zimmermann an idea. He met with

1. This is almost identical to what happened in 2010/11 to several computer security programmers including David House, Jacob Appelbaum and Moxie Marlinspike. In all three cases, US Customs and Border Patrol agents denied them access to a lawyer and seized all their electronic equipment without a warrant. They were interrogated about their cryptographic work and possible association with WikiLeaks.

the editor of MIT Press, which published technical books around the world. Would MIT publish a PGP users' manual that included the entire source code printed in machine-readable font? The editor agreed to a small print run, the book was shipped everywhere in the world, people tore off the covers and scanned in the book and – *voilà!* – there were copies of PGP all around the globe – without breaking any laws.

'This blew a hole a mile wide in the US export control regime for cryptographic source code,' Zimmermann said. After that the government's case crumbled and in 2000 the controls ended. 'We beat them. Sometimes you just win and that's what we did.'

After September 11, 2001, Admiral John Poindexter, former US National Security Advisor under President Ronald Reagan, and Brian Hicks from Science Applications International Corporation (SAIC) went to the US Department of Defense with a proposal to combat terrorism. It was called Total Information Awareness.

The idea involved the government sucking down every piece of digital data it could find on people: Web browsing history, emails, all medical records, credit card transactions, phone bills, etc. All without a warrant or probable cause. The goal wasn't to collect it for any specific purpose, rather to add it to a giant government database which officials could then analyse in vast fishing expeditions, trawling the entire population for suspicious activity. The project also set aside money to fund emerging biometric and other surveillance technologies.

Poindexter himself had a dubious pedigree when it came to national security. It was his idea to secretly sell missiles to Iran in order to pay hostage ransoms, and then funnel the profits to illegally support the contras in Nicaragua, in what became the scandal known as the Iran–Contra Affair. He'd been convicted in 1990 on five felony counts of misleading Congress and making false statements, but was let off by the appeals court because he'd been granted immunity for his testimony by Congress. Some might think this is exactly the sort of behaviour that is a threat to national security but instead Poindexter became the head of the Information Awareness Office in the Defense Advanced Research Projects Agency (DARPA)[2] and set about building an enormous secret state database that would snoop on every American. The public first learned about the project in a February 2002 article in the *New York Times*. There were few details apart from an announcement that DARPA's budget would be increasing dramatically the following year to refocus on new technologies such as 'data mining'.

The project's overseers placed public acceptance at such a low priority that the logo was literally the eye of God looking down from a pyramid with a light beam shining out.

2. DARPA financed the research that led to the creation of the Internet.

The Latin for 'Knowledge is power' was also inscribed on the logo. (At least you couldn't say they were hiding their true intentions.) As details of the project emerged, Americans grew more concerned, and in November 2002 *New York Times* columnist William Safire wrote a scathing piece titled 'You Are a Suspect' in which he outlined some of what the $200 million database would hold on 300 million Americans:

> Every purchase you make with a credit card, every magazine subscription you buy and medical prescription you fill, every Web site you visit and e-mail you send or receive, every academic grade you receive, every bank deposit you make, every trip you book and every event you attend – all these transactions and communications will go into what the Defense Department describes as 'a virtual, centralized grand database.'
>
> To this computerized dossier on your private life from commercial sources, add every piece of information that government has about you – passport application, driver's license and bridge toll records, judicial and divorce records, complaints from nosy neighbors to the F.B.I., your lifetime paper trail plus the latest hidden camera surveillance . . . [3]

To counter the bad publicity, the Information Awareness Office renamed the project *Terrorism* Information Awareness, but the public weren't fooled. In September 2003, Congress voted to strip the project of its funding and TIA was no more. At least that's what the American people thought. While its funding was cut, some elements of TIA carried on within other government agencies. Congress stipulated that those technologies were to be used only for military or foreign

3. http://tinyurl.com/cozfs6

intelligence purposes and not against US citizens but as I'll discuss in Chapter 7, the amount of data collected on all of us has never been greater.

'It seemed like we'd won'

After winning two big battles like this against the US government, privacy campaigners and cypherpunks could be forgiven for believing they'd won the info war. But if this was the case, why do we live in a world where surveillance is at an all-time high and individual privacy at a new low? One answer lies in the fact that the digitisation of information can both set us free but also imprison us. It means we are in danger of being recorded everywhere we go, all our actions stored on a database, our privacy lost to nosy state officials and private companies in a great Panopticlick of which Jeremy Bentham could only dream.

I met up with original cypherpunk John Gilmore to get his thoughts on the current situation. Gilmore is a legend in computing circles. He's in his fifties and uses the wealth he made as one of the first employees of Sun Microsystems and founder of Cygnus Solutions to campaign against government surveillance. Gilmore founded the Electronic Frontier Foundation in 1990 with Mitch Kapor, former president of Lotus Development Corporation, and John Perry Barlow, a Wyoming cattle rancher and lyricist for the Grateful Dead. They were part of an electronic community called the Whole Earth 'Lectronic Link (now WELL.com) and were the first to warn of the civil liberty issues arising from digitisation and the Internet. Gilmore is a regular at the Burning Man festival and when we meet at the Velo Rouge Cafe in the Haight-Ashbury district of San Francisco he looks exactly as you'd expect a genius eccentric developer to look: long hair, T-shirt, glasses and a calm, almost Buddha-like demeanour.

'We won half the battle and lost half and got lazy,' he tells me. 'What we won in the United States was the legal right to publish cryptographic software but what we didn't do was actually build it into the infrastructure, we didn't embed it. Having the right to deploy encryption doesn't mean that in your everyday life it is going to get used. Most emails go un-encrypted over the Internet as does most Web traffic. Your cellphone has crappy encryption. The reason it's crappy – I know the guy from Motorola who designed it, he's a crappy cryptographer and the committee allowed his crappy design to become the standard encryption for the North American cellphone standard. Governments want it that way. The last thing they want is end-to-end encryption. They want to squeeze the guys in the middle to get at what you're saying. We have two or three places where it's well embedded but that's about it.'

The cypherpunks understood where the Internet was headed and how important digital communication would be in the future. They could see the potential for mass government surveillance and wanted to ensure citizens' privacy was protected. The cypherpunks prevailed at the time, but those pushing for mass surveillance weren't put off. They simply changed tactics. They'd learned a few things from their failures. Firstly they learned to do better public relations. No more all-seeing eyes. Secondly, they learned to go where the data was rather than trying to store it all themselves.

If I were to ask a room full of people, 'How many of you would like to wear a government-mandated tracking device at all times?', I imagine few would raise their hands. But if I were then to ask the same group, 'How many of you are carrying mobile phones?', then the answer would probably be all of them. A mobile phone is just a tracking device that transmits your real-time position to one of a few telecommunications companies. What few people realise is that companies such as Google and Apple harvest this data and all are required to provide it to governments when legally requested.

One man who knows all too well the dangers of mobile phone tracking is Isa Saharkhiz, a reformist, pro-democracy journalist in Iran who has long lobbied for press freedom in that country – exactly the sort of person Hillary Clinton praised in her Net freedom speech. He was former Director of Media Relations at the Ministry of Culture during President Khatami's administration and is notable for loosening censorship and encouraging democratic reform. The 'Tehran Spring' came to an end in the late nineties when hardliners regained control of Iran and began shutting down newspapers. Saharkhiz responded by founding a progressive monthly called *Aftab* (Sunshine) in December 2000.

In June 2004 that paper was shut under the pretext that it contained insults to the Supreme Leader Ayatollah Ruhollah Khomeini. Perhaps not coincidentally this was a month after Saharkhiz had debated with the hardline cleric Gholam Hossein Mohseni-Ejehei at a meeting of the council that monitors the Iranian press. The cleric was angry over an article in another paper about relations between the sexes, and he made disparaging remarks about the reformists. Saharkhiz objected to these criticisms and challenged the cleric to a public poll to see who was more popular among the people, the hardliners or the reformists. Ejehei reportedly reacted by throwing objects at Saharkhiz and biting him. Saharkhiz tried to take the case to court but it was not prosecuted.

Saharkhiz continued writing pro-democracy pieces in other publications but two weeks after the disputed re-election of President Mahmoud Ahmadinejad in 2009, he was arrested on charges of 'insulting the Supreme Leader and propaganda against the regime'. He was beaten and tortured. In September 2009 he was sentenced to three years in prison, a five-year ban from journalistic activities and a one-year ban from travel abroad.

One of the things Saharkhiz discovered while in Tehran's Evin prison, with his ribs broken by police, was how Iranian security officials had discovered his whereabouts and listened to his telephone conversations thanks to interception technology supplied to them by Nokia Siemens Networks. The network had 'back doors' built in so the government could easily intercept the communications of the opposition movement and anyone else who was a perceived threat to their authority. The back doors allowed agents to identify, target, monitor and track down protestors who were then put in jail, beaten and sometimes killed.

Saharkhiz and his son, Mehdi Saharkhiz, filed a lawsuit in US Federal Court in Alexandria, Virginia, against Nokia Siemens Networks and its parent companies Siemens AG and Nokia Inc, alleging human rights violations committed by the Iranian government through the aid of spying centres which were provided by Nokia Siemens Networks. They claimed Saharkhiz's arrest, torture and detainment were a result of the surveillance and monitoring of his mobile phone communications. Nokia Siemens Networks admitted selling a monitoring system to the Iranian regime in 2008 but said the Lawful Interception Management System was simply a standard feature of the network and that any business selling networks was also intrinsically selling the capability to intercept any communication that ran on the network. This was not a mandate from Ahmadinejad or Khomeini. It was a mandate from the US government.

Saharkhiz could be monitored because all modern mobile communications networks include an interception capability as specified by US law and European trade standards. To understand why, we need to go back to 1994, during Bill Clinton's presidency, when Congress passed the Communications Assistance for Law Enforcement Act (CALEA). The FBI pressed for CALEA because telephone companies were switching to digital networks. The agency claimed traditional methods for eavesdropping, which involved physically tapping the line, were no longer effective. Police and security officials argued that they needed surveillance capability built into the digital networks to catch criminals.

This policy concerned the telecommunications industry in the US because it would be hugely expensive. In addition, it would lead to inferior products by design (they mandated building in a vulnerability), since the companies making these products would have to compete with those from European companies that didn't include back doors. The Clinton administration responded by setting aside $500 million of public money to help companies redesign and deploy the wiretap networking equipment. Additionally the government lobbied Europe and succeeded in persuading the European Telecommunications Standards Institute to make lawful intercept capacity a standard within all telecommunications kit across Europe.

As a result of CALEA's passage, all of the world's major telecommunications companies now have interception 'back doors' built into their telephony technology as standard. That's why Nokia's system automatically included the interception capacity which helped the autocrats in Iran spy on Saharkhiz.

It's also why the intelligence services in Egypt, Tunisia and Libya could easily track down and arrest so many democracy protestors in those countries during the uprisings of 2011 – and anyone else who challenged their authority.

A Wiretappers' Ball

It used to be that running a police state required a tremendous outlay of resources, from hiring informants to the central collection and storage of mountains of paper files. As we move our lives onto digital networks, we create a handy one-stop shop for the nosy official. There's no need to hire informants – they can simply eavesdrop on all our communications. They don't need to store the data centrally, they just go to where it's collected: social networks and telecoms companies. By outsourcing surveillance gathering to private companies, the state gains a whole new level of efficiency.

The US, in principle, boasts some of the strongest legal protections for the individual against the power of the state. In practice, however, these rights aren't as strong as they initially appear.

Strictly speaking, wiretaps require a warrant which can only be given by a judge according to strict rules, but there are many ways the government can monitor an individual without the need for this warrant. 'Pen register' court orders are one example. A pen register was originally a device that records numbers dialled from a telephone, but the term has been extended to include any technology or software that can similarly collect information about the other phone lines that are communicating with a surveillance target. Pen register orders can be obtained simply by certifying to a judge that the desired information is relevant to an ongoing criminal investigation. While interception (wiretap) orders are required to monitor the contents of real-time communications, the non-content information sought via pen register orders, which includes location, to/from and other transactional data, can be obtained much more easily. (At a state level, it's a different story – there are state law enforcement agencies that conduct mobile phone location surveillance with no court order at all.)

The FBI can obtain signalling information (details of all incoming and outgoing calls, for example) at the touch of a button with the Digital Collection System Network (DCSNet). This connects FBI wiretapping rooms to switches controlled by private landline operators, Internet telephony providers and mobile phone companies. Through the DCSNet the FBI can tap regular landlines, mobile phones, SMS and push-to-talk systems anywhere in the United States.[4]

Other law enforcement agencies might look enviously on the FBI's ability to eavesdrop with little regard for constitutional protections, but they, too, are hitting telecom companies hard. Sprint is just one of the service providers who attend the annual surveillance industry conference in Washington, DC, the Intelligence Support Systems for Lawful Interception, Criminal Investigations and Intelligence Gathering – also known as the Wiretappers' Ball.

One of the keynote speeches in 2009 was given by Sprint Nextel's Manager of Electronic Surveillance Paul Taylor, who complained, revealingly, that the company was getting hit with so many law enforcement requests for its customers' data that it was finding it difficult to cope:

'[M]y major concern is the volume of requests. We have a lot of things that are automated but that's just scratching the

---

4. Steven Bellovin, a Columbia University computer science professor and long-time surveillance expert, in *Wired* magazine: http://tinyurl.com/2j8g4t

surface. One of the things, like with our GPS tool. We turned it on the Web interface for law enforcement about one year ago last month, and we just passed *8 million requests*. So there is no way on earth my team could have handled 8 million requests from law enforcement, just for GPS alone. So the tool has just really caught on fire with law enforcement. They also love that it is extremely inexpensive to operate and easy . . . I just don't know how we'll handle the millions and millions of requests that are going to come in.'

Mr Taylor told the conference delegates that his electronic surveillance group at Sprint comprised three supervisors, thirty technical support staff and fifteen contractors. Dealing with subpoenas (which are needed for stored content, stored records and anything historical) there were an additional thirty-five employees, four to five supervisors and another thirty contractors. So in just one US telephone company that was over 110 people working to comply with government requests for its customers' records even with many services being automated.[5]

The US government initially tried to gather geographic (geo)data under a pen register order but the courts ruled

5. Christopher Soghoian, a DC-based academic specialising in online privacy and government surveillance for his PhD research, quoted the above material in an article designed to alert the public to these statistics. The company denied the figures, but fortunately Soghoian had recorded the talk and posted an mp3 recording of it on his website. A couple of days later he had a phone call from an executive at TeleStrategies, the organiser of the ISS World conference, claiming Soghoian's recordings violated copyright law and would need to be removed. He didn't agree but in a spirit of goodwill removed the recordings but left the post online. There are laws requiring disclosure of the number of warrants issued for state and federal surveillance but the Sprint figures were not included. Soghoian says the Department of Justice routinely ignores the law, and goes many years without providing the legally mandated reports to Congress.

there was a greater expectation for location data to be private.

In 2005, however, the Electronic Frontier Foundation learned that the government was routinely seeking, and getting approval for, real-time phone tracking data under a law governing historical data (the Stored Communications Act). When the EFF exposed this activity the government claimed to have created a new type of 'hybrid' court order that was a combination of pen registers and stored communications. While neither of the orders gave them what they wanted, by combining them the government claimed that the phone tracking was legal. 'It's like adding up zero and zero and saying it equals one,' said Kevin Bankston, the EFF's senior lawyer. 'We've been fighting this. We've had a lot of judges coming to us who have been getting these applications asking us to brief them on it.'

The EFF has had a tough time getting a case to court to challenge the government's strategy. Most of the orders are sealed and whenever the EFF has tried to challenge the unwarranted phone tracking in court, the government drops the case and goes to a different judge.[6]

Internet interception

I asked Kevin Bankston what the EFF did to stop the FBI's push for government-mandated interception. He told me EFF lobbied aggressively against CALEA but back in the nineties the main concern was keeping the Internet free from state surveillance. To ensure this, the EFF's executive director at the time, Jerry Berman, agreed to a compromise with the US government known as

6. This warrantless telephone tapping burst into the public realm in a much-publicised scandal in 2006, when the NSA was discovered to have secretly collected billions of phone call details from regular Americans, ostensibly to fight terrorism.

'hands off the Internet'. EFF would accept the need for lawful interceptions of telephone networks as long as it was not applied to information services, which is what the Internet was called back then. They argued that mandating interception technology for the burgeoning Internet industry would kill it. The Clinton administration agreed and so the law excluded Internet and information services but went into full effect on telephones. Many of EFF's board members and supporters were upset that such a compromise had been made, so much so that the organisation split and the HQ moved to its current location in San Francisco. Berman stayed in DC where he started another organisation called the Center for Democracy and Technology.

The compromise became even more distasteful when it was clear the government had no intention of keeping its side of the bargain. Almost as soon as CALEA was passed, the FBI and Department of Justice began lobbying to extend it to the Internet. They finally succeeded in 2005, when the FBI petitioned the Federal Communications Commission (FCC), in charge of administering CALEA, to interpret the law as extending beyond the phone network into broadband and VoIP (Voice over IP) providers that were interconnected with the phone network. The rationale was that traditional phone companies like AT&T were, not unfairly, asking why they had to build the government-mandated back doors into their telecommunications systems when Skype and other Internet-based providers did not. The FCC approved the extension.

Surveillance must always be done within the constraints of laws protecting citizens' rights. These laws exist for a reason: to act as a check on the very considerable power of the state. Once these checks are removed, the result is a concentration of power which has proven to be highly dangerous. There are laws already that permit Internet surveillance, regardless of what programs or protocols are used to communicate. What the FBI wants is tap-ability on demand, a way to bypass constitutional protections. Surveillance has actually become much easier in the digital age: agents can tap mobile phones, access reams of electronic data such as email, conduct DNA identification tests and track people's locations using mobile phone signals. For our own protection it would be better to grant the courts more power over state surveillance, not less, to ensure this increased spying ability isn't abused.

The USA Patriot Act does precisely the opposite. Signed by President Bush on 26 October 2001 after the World Trade Center attack, this law expanded the US government's ability to conduct searches and surveillance on its citizens while eliminating many of the checks and balances provided by the courts. Under the Patriot Act, the need to show probable cause was removed so government can get a court order simply by stating that it is their belief the information sought will be relevant to a criminal investigation. The court has to grant the order.[7] The court cannot consider whether the information sought would be relevant. 'It is incredibly easy for the government to get and yet, particularly in the Internet context, incredibly powerful,' Bankston told me. 'Imagine how revealing the logs of every email and IM [instant message] can be.'

Unwarranted search and seizure

The Fourth Amendment protects American citizens against unreasonable intrusions by mandating that a warrant is needed,

7. In 2011 the US Department of Justice got a court order for the Twitter data of Rop Gonggrijp, Birgitta Jónsdóttir, Jacob Appelbaum, Julian Assange and others, showing that it's not only Americans who are affected by these laws.

signed by a judge and showing probable cause – i.e. there are to be no fishing expeditions by police or state agents. The amendment was included to protect against the abuse of writs of assistance that colonialists had suffered at the hands of British state officials. Writs were general warrants issued by the British Parliament to allow customs officials to search for smuggled goods. But in the American colonies they were used by agents of the British state to interrogate people and raid their homes on the pretext of searching and seizing any 'prohibited and uncustomed goods', which often meant 'seditious' publications that criticised government policies or the King. The Colony of Massachusetts passed a law banning these general warrants in 1756 but it was overturned by the governor for being contrary to British law and parliamentary sovereignty. This decision helped spark the American Revolution. It's therefore ironic that the Patriot Act does so much to strip away Fourth Amendment protections and grant to US officials many of the same unlimited powers of search and seizure that so aggravated the American revolutionaries.

A big problem for the courts has been reinterpreting constitutional rights in the face of rapidly changing technology. In the old days people stored their most personal or precious possessions in their houses so it made sense to focus on protecting the home. However, we're now more likely to store private information electronically with a third party.

It took about forty years for the Supreme Court to rule that tapping telephones was similar to a physical search and required a warrant based on probable cause. Initially, telephone calls were routed via semi-open networks where the operator could (and often did) listen into calls. As such, it was thought callers couldn't have a legitimate expectation of privacy. It was only in 1967 that the Supreme Court ruled that the content of telephone calls was protected.[8]

The privacy law surrounding our emails is similarly outdated, based on the technology of the first email services of the 1980s. Back then, people dialled up their provider to download email onto their home computer. Mail left for over 180 days was considered in storage so was not subject to the wiretap protections which were for information in transmission. This means email older than 180 days doesn't require a warrant whereas anything newer does. Now, with cloud services and extensive storage available through services such as Gmail, our primary archive of email is held more or less indefinitely. Ironically, this means the most important or sensitive emails receive the lowest legal protections. The law is also weighted to protect unread mail over read mail so, strangely, spam that remains unopened because it goes straight to your junk folder has more privacy protections than read mail in your archives.

If we go back to the data the government hoped to collect in Total Information Awareness it's not dissimilar to what is gathered by companies like Facebook and Google. In addition there is an enormous amount of communication that once went over copper wires that is now in the air and can be grabbed by anyone with an antenna. WikiLeaks, for example, published an archive of every pager or text message sent in Manhattan on 9/11. It meant that some organisation (presumably the National Security Agency) was monitoring all such communication before 9/11 even happened.

The FBI and Department of Justice frequently lobby Congress to increase their surveillance powers. In 2011 the goal

8. Katz v. U.S. (389 US 347) 1967.

was to expand surveillance back doors to all Internet communications, including Facebook, Google, peer to peer messaging services, and encrypted communications such as BlackBerry email. Rather than setting an example for the world on citizens' rights against the tyranny of the state, the US government is taking its lead on Internet control from repressive regimes around the world, like the the United Arab Emirates, which bans individuals and small businesses from using secure encryption that cannot be monitored by the government. The 'crypto wars' are back and the stakes are bigger than ever.

The law of unintended consequences arises again over the extension of CALEA to the Internet. According to Bankston, the EFF learned from sources inside Google that the Chinese government hacked into their systems using the lawful intercept capabilities imposed by the US. The governments of the United Arab Emirates and Saudi Arabia say their restrictions on encrypted BlackBerries, mobile phones and the Internet are no different from the 'lawful intercept' capacity mandated by the US government.

What is notable about the crypto wars of the nineties is that the debate about national security was held in the open and everyone participated, including the White House, the NSA, the FBI, the courts, Congress, the computer industry, civilian academia, the press, the police and the people. Terrorism was discussed and it was one of the main concerns. Even so, the collective decision (over the NSA's objections, of course) was that on the whole, society was better off with encryption available to all, unencumbered by government back doors. The danger of allowing the police, intelligence agents and security services to decide policy is that they often ignore the benefits of freedom, focusing instead on the risks. The more we are afraid, the more likely we are to cede our liberties and hand power over to the state.

Filtering out free speech by design

The Open Net Initiative, which tracks Internet censorship globally, has shown that since 2002 the number of countries censoring content on the Web has increased from four to forty. Some of the biggest companies on the Internet are now wondering how to protect their users from an onslaught of requests from police and state authorities. Google is coming under increasing pressure to hand over data on its customers and remove pages from its global index. In Google's London office, Scott Rubin, head of public policy strategy for Google in Europe, Africa and the Middle East, showed me a transparency application that lists the number of requests Google receives from governments around the world for user data and page removals. The top-scoring countries include India, Germany, Brazil, the UK and the USA. There are no figures for China as Chinese officials consider such demands to be state secrets – even the censorship is censored. 'Censorship is growing overall,' says Rubin, 'and we see that as a growing problem.' Google released its application in hopes that publicity might decrease the number of requests that aren't legitimate.

It isn't just China. A number of liberal democracies are introducing filters, surveillance and Internet blacklists and even kill switches that grant the government the ability to shut off the Internet entirely.

Internet service providers are compelled to use blacklists for filtering certain Internet material. Australian broadband forum Whirlpool faced fines of $11,000 per day for posting a link to a blacklisted anti-abortion website. WikiLeaks was on the

Australian blacklist in 2009, though once this was made public (through a leak that WikiLeaks itself published) it was restored. Australia's Communications and Media Authority orders the censorship of sites but as a department it is exempt from freedom of information requests, meaning there is no way the public can check to see which sites are blocked. When WikiLeaks obtained the secret blacklist it found the Australian government had understated the number of banned webpages by more than a thousand. Among the 'dangerous' sites were two bus companies' webpages, online poker sites, Wikipedia entries, Google and Yahoo group pages, a dental surgery and a tour operator.

The Australian blacklist forms the basis of the government's proposed mandatory ISP-level Internet censorship legislation. Many countries – including Britain – now use filtering systems to restrict Internet access to outlawed material. In the UK the Internet Watch Foundation keeps a secret list of websites that ISPs must block; in Germany and Canada ISPs use similar blocking tactics; in Italy gambling sites are blocked. There is undoubtedly content online that is illegal, and by all means go after the owners of these sites and prosecute them. If that is not possible then we need to know why it's not possible and think about changing the laws. But those in power can all too easily use secret blacklists to silence opponents. There is every opportunity to block politically inconvenient or embarrassing websites that challenge authority.

The Internet decentralises power to the extent that it enables free expression and free association, but across the world there is a growing trend by authorities in both democratic and repressive governments to seek the means of controlling these networks, to use them to monitor all our speech and association in an efficient, centralised way. The great challenge in the digital age will be finding a way to harness the democratic effects of digital technology while checking its authoritarian tendencies.

On 16 June 2010 one small country took up this challenge with vigour. Icelanders had seen the worst and were ready to take a leap of faith. They were going to hitch their future not to the authoritarian desire to control information but to setting it free.

Since February, Birgitta Jónsdóttir, Smári McCarthy and Herbert Snorrason had been working in and out of committee rooms to get the Icelandic Modern Media Initiative (IMMI) onto the floor of Parliament for a full vote. No one thought they had a chance but after months of lobbying, a rare cross-party consensus developed around the resolution with fourteen MPs pledging to speak on it (at least two from every party) when it came up for a vote. The vote came on the last day of the parliamentary session on 15/16 June.

There is always a backlog of bills to get through and the night of the 15th was no exception. Smári watched from the public gallery until 1 a.m., then joined a crowd of parliamentarians at the nearest bar to get a last beer in before closing time. It wasn't until 4 a.m. that Birgitta rose to speak for the resolution. Despite the hour, the chamber was full this time. She told her fellow MPs they'd seen the dangers of an unfree press, and reminded them of the gagging of RUV's nightly news by Kaupthing Bank. The potential for IMMI was already clear: important newswires and human rights organisations had moved to Stockholm on the strength of the Swedish Press Freedom Act. Similarly, *Malaysia Today* had relocated to the United States after persecution in its own country. 'The world is looking for an internally consistent set of rules that place clear limits on the risks faced by publishers,' she said.

Then it was time to vote. Birgitta wondered, 'Is this going to pass?' She knew they had support. But how much? She watched the wall where the votes are tabulated. A yellow dot went up – an abstention, then a green 'yes' vote. Then another green and another and then the wall was awash with green dots. Across all parties the Icelandic MPs voted unanimously in support of making their country a new haven for free speech. With such a clear statement from Parliament, the government was tasked to craft and implement the thirteen new laws needed to make IMMI a reality. The date set for completion was summer 2011. Birgitta stayed on until six in the morning. She didn't go out and celebrate. Instead she hurried home to get online and let all the supporters around the world know the momentous news.

7

Private Lives

London, July

Bradley Manning spent the summer in a cell in Kuwait's Camp Arifjan awaiting charges. They came on 5 July: two under the Uniform Code of Military Justice for leaking the 'Collateral Murder' video and US State Department cables. The first charge involved eight counts for alleged criminal offences involving communicating national defence information to an unauthorised source that brought 'discredit on the armed forces'. The second included four non-criminal violations for transferring classified information from a secure network to his personal computer. If found guilty of all charges, he would spend the next fifty-two years in prison.

US government officials were keen to speak to Julian Assange about his role in the leaks but he was now in hiding.

'We'd like to know where he is. We'd like his cooperation in this,' a US official told the press in June. Investigators from the Pentagon were trying to track down the WikiLeaks frontman after Manning apparently 'confessed' to a hacker that he'd leaked not only the 'Collateral Murder' video to

Assange but also a number of other classified records including 260,000 State Department diplomatic cables.

The State Department were looking for Assange to convince him not to publish these cables if he had them, arguing it would damage not only national security but also diplomatic efforts around the world. Assange issued a denial, claiming allegations that 'we have been sent 260,000 classified US embassy cables are, as far as we can tell, incorrect'. He was scheduled to speak at two conferences in America in early June but dropped out of one and Skyped in from Australia for the other. Daniel Ellsberg, the Pentagon Papers whistle-blower, told Assange he was better off not coming back to America, as the government's interest 'puts his well-being, his physical life, in some danger now'.

In Britain, *Guardian* journalist Nick Davies was also trying to track down Assange. He'd read the chat logs published in *Wired* magazine and if Julian Assange was in possession of a quarter of a million US diplomatic cables then Nick wanted to talk to him. Such a trove of material could be the biggest story of the decade. But first he had to find Assange, who was even more elusive than usual due to Ellsberg's warning. Nick compiled a list of Assange's associates and began phoning and emailing, asking them to pass on a message to the lead man: 'This is the biggest story on the planet and no one is covering it. I want to. Mainstream media coverage is what Julian needs now more than ever. It's important political protection. And I can give him that.' A few weeks later, one of Assange's colleagues phoned to say Julian would be speaking in Brussels on 21 June.

'I got on the first Eurostar I could,' Nick said. 'In the meantime I phoned the *Guardian*'s Brussels correspondent Ian

Traynor and told him it was absolutely imperative that he get hold of Julian Assange at the conference.'

Traynor was able to persuade Julian to meet with Nick the following day.

'Julian said he had 2 million documents he was preparing to put on the Internet,' Nick continued. 'I was wondering, "How do I get him to not put it on the Internet and let me write stories about it?"' He'd formulated a plan and laid it out, going through all of Julian's options as he saw them. Assange was in a precarious position and needed protection. The US was unlikely, despite what Ellsberg said, to physically attack Assange. They could launch a legal or technical attack but both would take time and meanwhile the data could be published instantly around the world. That left the most likely option as Nick saw it – that the US would launch an 'information war' against WikiLeaks and Assange. 'Media manipulation – that's where the US will attack you. It's where they attacked on the video.' A couple of years earlier Nick had published a book, *Flat Earth News*, in which he'd investigated, among other things, the US government's increasing use of 'media management' as a tactical resource, particularly in the Department of Defense and intelligence agencies. That tallied with Julian's feeling that the American press had 'taken the Pentagon's line' on the 'Collateral Murder' video with the debate quickly shifting from the behaviour of the troops in the video to that of WikiLeaks in publishing it.

'The first thing you need to do is to get yourself so high on the moral high ground that you can't be attacked,' Nick advised. A combination of the *New York Times* and the *Guardian* would grant huge doses of political immunity. The *Guardian* wanted the *New York Times* on board as protection against the government stopping publication with a court injunction (a problem in Britain, where there is not an equivalent to the First Amendment to protect free speech). It was unlikely President Obama would try to get a court order against the *New York Times*.

They sat in the café for six hours discussing the details, and finally ended with a deal: four packages to include US military reports from Afghanistan and Iraq, the US diplomatic cables and the personal files of Guantánamo detainees. They talked about another video Assange had of an American air strike in Granai, Afghanistan, but Julian thought the footage was confused and did not tell a story like the Apache helicopter video. 'The one thing he wanted was a voice in the timing,' Nick told me. 'If the US attacked him, he would put everything online.'

What Nick needed now was access to the material. Julian proposed creating a new website, posting some of the raw data to it and giving Nick a limited time in which he could download the data. That's when Julian took a paper napkin from the table, joined up the letters of the name of the café and used it to create a username and password. 'He did that on two napkins and we each put one in our pockets.' Assange wasn't going to give him all the material at once. He would start with the 91,000 US military reports from the war in Afghanistan. Nick got back on the Eurostar and headed to London to discuss with *Guardian* editor Alan Rusbridger and investigations editor David Leigh how to proceed.

I was looking for Julian, too. He'd told me at the Norway conference back in March that he only spoke on cryptographic phones. Since the conference we'd exchanged a few emails.

I was still full of admiration for this warrior for freedom of information; his fearlessness had impressed me a great deal. I'd written that I'd felt held back by the nebulous concept of 'getting into trouble'. He'd responded in florid terms:

> Dearest Heather, courage is not the absense [*sic*] of fear, courage is the mastery of fear. We fly in airoplanes [*sic*] every month, machines of abject terror. But through couragous [*sic*] examples we are able to see through our fears, and go ahead anyway, not because we want to throw our lives from the sky, but because we understand our fears illusions [*sic*], and that to not fly is to take on the greater risk of living a life unlived . . .

Yet he remained elusive, even more so once Manning was arrested. I would get random, intriguing messages:

> Lovely heather, I'm fine. But lots of backroom action at the moment . . . let us do meet, some sunny day.

Eventually, I ran out of time and asked him bluntly if he wanted me to write about him or not. He responded in typical style:

> I will have you, Heather, of course I will. But let us be messiahs to generation WHY, not a bunch of aging hacks looking for a pension . . . regards from intrigue hotel . . . I have more interesting adventures for you . . .

As it happened Assange sprang on me very much like the Scarlet Pimpernel. I'd been in America for a few weeks and he'd asked when I would be back in London. A few hours after I'd landed he called. He was in London, he said, and wanted to stay with me.

'I have a fever. I'm not sure yet if it's going up or down,' he tells me. 'I need some mothering. Someone to make me chicken noodle soup and bring me cookies in bed.'

This isn't quite how I'd imagined our interview taking place, so I politely explain that's not really my scene.

'Don't you have a maternal instinct?' he asks in a breathy, slightly hurt tone.

'No, sorry, I don't, particularly not for grown men.' I pause. 'Nurturing isn't something I'm into.'

'Really? How intriguing. You don't feel compelled to nurse me back to health?'

'No.'

'Are you really so hard-hearted, Heather? Don't you have a softer side?'

'I'm prickly, I suppose.' I wonder where he's going with this. 'Some people, once they get to know me, might say I have a softer side underneath all those prickles . . .'

'And are you soft-hearted underneath?'

'Perhaps.'

I definitely want to interview Julian and having him in my flat would be a unique opportunity to do so, but I'm wary. For one thing I get the sense he'd be difficult to shift once ensconced.

'There's also the issue of my husband,' I say, fairly certain he would not be at all keen to find Julian Assange in the spare bedroom. There's a moment of silence while this new piece of data is integrated into whatever calculations he'd made.

'I'll find somewhere else,' he says.

'Are you sure? If you can't find anywhere then I'll see what I can do. I don't want to leave you on the streets. But let's definitely meet. I need to talk to you.'

Later in the day, after a meeting he has with the *Guardian*, we speak again.

'Hello, Heather.' It's that deadly sonorous voice again.

'How was your meeting at the *Guardian*?'

'Oh, OK. Pretty good.' He sighs. 'I just have so much to do.'

'Yeah, it's a tough life being a messiah.'

'Will you be my Mary Magdalene, Heather? And bathe my feet at the cross?'

This is a new one on me. What does a person say to such a question? At that time I did genuinely like Julian. When I'd met him at the conference he was like a bolt of lightning. But even so – foot-bathing?

I'd reached a point with Julian where the personal and the professional had begun to blur. He's the world's most famous leaker; I'm a freedom of information campaigner, so we've a lot to talk about. But he was unsettlingly, even bafflingly, unaware of any notion of personal boundaries. The evening after I'd refused him a room at my flat I went over to a mutual friend's house where he was staying and he recounted to us several stories of various female conquests. When I asked for a private word to discuss my research he saw this as an opportunity to try and make another, and I couldn't have felt less comfortable alone in that room with him. Walking along a Georgian street to Pimlico tube station that night, he stopped to say he could never understand how people could bear to live in London, 'all these houses, everyone on top of each other', staring at me as he said this, pressing me against a brick wall, in full view of the city around us.

Privacy in the age of identification

Perhaps a basic appreciation of personal boundaries is too much to expect from someone who first made his name as an international hacker, but my latest encounter with Julian got me thinking about whether 'privacy' really means anything in the digital age. This blurring of boundaries between private and public, friends and strangers, is now common, as social networks and interactive communication have dismantled traditional social norms. Whether we view that as good or bad is a matter of personal choice but we should at least be aware how these spheres have changed. Social networks have spread around the globe without anyone giving much thought about how we, personally, are affected. We can see the benefits clearly enough, otherwise we wouldn't sign up, but there are two areas where there are considerable costs: to our identity and privacy. If we don't start giving some thought to how we define privacy and what sort of privacy we want online, then we risk letting governments and corporations define these issues based on their own interests – power in the case of governments, profit for companies.

Whereas governments and corporations have the resources to 'disappear' items they don't like from the Internet, most of us don't, and anything posted on YouTube, Twitter or Facebook is up there pretty much permanently. Private individuals can suddenly find themselves at the centre of worldwide attention for both good and bad reasons. The hive-mind group of hackers 'Anonymous' uses 'doxing' as one of its main weapons: dredging up from the Internet all sorts of personal information – whether photos, documents or videos – that most of us thought was ephemeral, or at least safe, and publishing it for mass harassment as it whirls round the world in a matter of seconds. Individuals who are targeted struggle to deal with the fallout.

The fact that young people are putting so much personal information online has led some to believe they don't care about privacy. I don't think that's the case. Rather we are having to learn a new set of values around privacy. People will always have secrets and even young people know how to keep them. The change is happening in the private sphere: something that is not a secret but at the same time not something we want widely known. Most of us speak differently when we're in an intimate 'private' space than in public – it's the difference we feel when we're speaking naturally among friends and then someone brings out a video camera. It's not that we have anything to hide, but we're more on guard, careful to moderate our language for the crowd. When politicians seem disingenuous it's usually because they speak all the time in this 'public' way. Much of what we once considered 'private' is moving into the public domain – at least for individuals. We have private online chats which, at the click of a button, can be published for all the world to see. We tweet to our friends and yet the whole world can read our thoughts. We post photos on Facebook and suddenly they appear in a national newspaper. If I can quote my favourite tech writer Danny O'Brien again: 'On the Net, you have public, or you have secrets. The private intermediate sphere, with its careful buffering, is shattered. Emails are forwarded verbatim. IRC [chat] transcripts, with throwaway comments, are preserved for ever. You talk to your friends online, you talk to the world.'

So is privacy really dead, as Facebook founder Mark Zuckerberg claimed? Before we can answer that we need first to define exactly what we mean by privacy. I think the best definition is one given by Supreme Court Justice Louis Brandeis in 1928 as simply the right to be left alone.[1]

The right to be left alone in an interconnected online Web can seem difficult. Almost by default to be online or on any social network is to agree to be identified and to have the information we post available to everyone for ever. Do we accept that willingly or has this simply become the default position because it is the way companies and governments want it? Is there an alternative? Certainly those with power, like corporations, governments and rich individuals continue to demand levels of privacy no longer possible for the rest of society who opt to use the networked Internet. In England, the rich and powerful go to High Court judges who routinely grant super-injunctions that ban speech as a contempt of court. One notorious judge is even granting *contra mundum* orders gagging speech around the world. For the rest of society without money or access to an English judge, it may be that we will have to become more forgiving and understanding of each other with so much information floating around. We could find a growing divide between those who can afford the luxury of privacy and the bulk of people who can't.

In the previous chapter I gave some indication of the information governments collect on us by eavesdropping on telephone and Internet networks. But companies like Microsoft, Google, Facebook and Apple collect personal information about us for their own purposes, too, and we have an even

1. 'The makers of our Constitution undertook to secure conditions favorable to the pursuit of happiness,' Brandeis wrote. 'They sought to protect Americans in their beliefs, their thoughts, their emotions and their sensations. They conferred against the government, the right to be let alone – the most comprehensive of rights and the right most valued by civilized men.'

hazier idea about how much and with whom that information is being shared. While some efforts are now being made to give people power over their data – such as two recently passed California laws that require companies to inform customers of data breaches and limit the amount of data that a financial institution can share with others – such laws are few and far between.

Companies are hard at work identifying users and their online patterns for various purposes not always even known to them. Collect data first, think of uses later, is the ethos at many companies. A *Wall Street Journal* investigation in July 2010 found that the business of spying on Internet users online is one of the fastest growing, with the top fifty websites in the US installing on average sixty-four pieces of tracking technology onto the computers of visitors, usually with no warning. A dozen sites each installed more than a hundred. The amount of cookies and other spying technology that companies deploy to monitor their users is growing daily.

Part of the problem with privacy is that it is not a fixed concept. It is moveable and changes in relation to time, age, personality, culture. How do we decide what is an invasion and what is beneficial for society? It would be wrong to fetishise privacy – after all, anonymity is a relatively recent phenomenon, having only really existed for the last few of humanity's 800 lifetimes. For the bulk of human existence we've lived in small groups where people knew a lot about each other though not much about those outside their immediate group. In the digital age there is a global record of everyone online. And this digital record has strange properties: being permanent in some cases but entirely malleable in others, according to the power and resources expended.

* * *

To discover more about online privacy I begin by visiting Ben Laurie, one of the foremost computer security researchers and a privacy advocate. He's the creator of Apache-SSL, a secure Web server application, and a core developer of OpenSSL, the world's most widely used cryptographic library. He was also one of the advisory board members for WikiLeaks and reviewed the design for the website before it launched. He tells me a funny story about how he stopped Jimmy Wales, the creator of Wikipedia, from suing Julian.[2] He's also a full-time employee of Google, investigating security for them on a number of projects.

Laurie lives in Acton, a leafy west London suburb of tall brick houses. The gate to his house is overgrown with vines and looks like the entryway into the Secret Garden. Near the door two recycling bins overflow with wine bottles and Budweiser boxes (the proper Czech kind, not the American impostor). I ring the bell and he comes to the door barefoot, wearing black jeans frayed at the bottom with the de rigueur black T-shirt, printed with 'Usenix Security Symposium'. He looks like an eccentric professor, with thick black/grey hair, metal-framed glasses and very long fingernails. We go into the kitchen where a Bengal cat (he has three) jumps onto the large wooden table that's stacked with copies of *Scientific American* magazine, newspapers and a half-empty bottle of red wine.

His office is a large room filled with half a dozen laptops, as many monitors, a locked rack of servers as tall as I am, shelves of computer programming and cryptography books, various phones, a few remnant CDs (he's transferred all his music, including an entire collection of Bob Dylan, to a music

2. 'He [Wales] said that WikiLeaks had misappropriated his name and his mark, but I said, "They're good guys" and he said, "All right then" and left them alone.'

server) and tangles and tangles of wires that spill out every-where like the guts from some enormous techno-whale.

'No system can absolutely guarantee anonymity online,' Laurie says as we settle down at the kitchen table. 'In future people will have to pay to be anonymous. Privacy will be a luxury in the identification age. We are reaching a time where it is pretty much impossible to do anything without, in some way, identifying yourself on the Net.'

He says the inherent difficulty in the structure of the Internet is that you are identified by your entry point, by your IP (Internet Protocol) address, which is strongly linked to your physical location. It's trying to remove this physical locator that is the focus for most privacy-enhancing technologies rather than changing policies around the collection and use of data. Laurie's work around privacy is to get us back to a more 'natural state', he says. 'When I walk out on the street, people can see what I do, but I'm OK with that because most of them don't know who I am and most of them I'll never see again. The Internet does not have that property. I am strongly identified. In order to restore a more natural state you want to have anonymity and introduce identification as needed.'[3]

But WikiLeaks promises to protect its sources' anonymity and he oversaw the architecture, so how does that work?

'I think it's very dangerous sending things in to WikiLeaks, because as I said, anonymisation is really hard, so you have plenty of opportunities to fuck up and be identified.'

This surprises me. 'So submitting documents is not secure?' Laurie says no. His advice to anyone wanting to leak would

3. When I talk with John Gilmore, the legendary cypherpunk we met in the previous chapter, he agrees, saying that bolt-ons for privacy are far less effective then making products with 'privacy by design' or not collecting the data in the first place.

be first to be aware that secret documents are often water-marked or have metadata for identification purposes. 'So my first step might be to steal a copy from somebody else, then get somebody else to do the mailing because I don't want that to be me either, and I'd encrypt it and all sorts of other stuff, maybe transcribe it before I send it. Basically, if I was in a situation where my life would be in danger if I revealed this stuff, I would be thinking very carefully about sending it to WikiLeaks, because I don't think WikiLeaks can make strong guarantees about anonymity, just like nobody can. They make it as strong as they can, but to a large extent it's not the recipi-ent's problem, it's the sender's problem. It's the sender that identifies him- or herself, not the recipient.'

Is the Internet a force for social good or the ultimate surveil-lance device? He says it's increasingly true that the Internet is being used as a surveillance system but that's mainly because of how much more we're doing online, and this data can be used in increasingly sophisticated ways, such as logging users' searches for analysis.

Even when companies anonymise search logs for open research, there have been a number of scandals where identities were deduced, causing embarrassment for both the company and individuals. A classic example was in 2006 when AOL released the anonymised search records of 500,000 of its users. A few days later the *New York Times* identified user number 4417749 as sixty-two-year-old Thelma Arnold from Lilburn, Georgia, based on 300 search terms including 'landscapers in Lilburn, Ga', '60 single men', 'numb fingers' and 'dog that urinates on everything'.

AOL was highly embarrassed: the CTO of the company and the researchers responsible for sharing the data were all fired and the data was pulled from the Web. The same year US movie renter Netflix published an anonymised data set of over 100 million movie ratings made by 500,000 subscribers to the online DVD rental service and offered $1 million to anyone who could create an improved algorithm for recommendations. In just a few days, someone announced they had de-anonymised the data, by comparing the Netflix data against publicly available ratings on the Internet Movie Database (IMDB). More red faces.

Although few companies are now willing to share even anonymised data with the public, many still use it internally or sell it. Why does this matter? Think about all the search terms you use and what that might tell someone about you. Depending on the person's agenda this information could do you a lot of harm. It's not a secret per se – you're not trying to hide it, but it is private. What if these records fell into the hands of a health insurer and you've been Googling various medical ailments? In the US, librarians have a long tradition of protecting users' privacy and confidentiality in relation to books borrowed. They have fought and won court cases to uphold users' privacy, but the USA Patriot Act effectively dismantled this hard-won right to privacy when it gave government agents permission not only to obtain library records, but also secretly to monitor electronic communications and prohibit libraries and librarians from even informing users of this surveillance. Yet library records are nothing compared to the information stored by Google and Facebook and because it is digitised, it can be analysed in much more powerful ways.

Google has gone on record saying it anonymises search logs. I ask Laurie for the details of what Google knows about us as he was involved in the anonymisation process.

When you do a search, he tells me, your IP, a set of numbers that connects to the computer server and which gives your physical location, is logged, along with other details such as the time and what you searched for. After nine months Google goes through those logs and drops some of the data so that you're less identifiable. That's voluntary but there is talk of making it obligatory, at least in Europe. Article 29, a working group set up as an independent European advisory body on data protection and privacy, wanted Google to drop search logs after one minute, which Laurie describes as 'an insane thing to ask, because it would prevent development of the search engine. If Google agreed, its service would get crappier over time and somebody else would take over as the best search engine, and then of course that company wouldn't agree to delete logs because to be the best search engine you've got to have logs.' Logs are necessary in order to diagnose and improve, and they have to identify people as much as possible – at least in the short term – in order to be able to do those things. 'If you completely anonymise then you lose so much of the benefit of gathering data that you might as well not. As far as we know it's impossible to completely anonymise – the only way to completely anonymise is to delete all of the data.'

There are other ways we are identified online: a whole selection of settings on our computers that websites harvest. These can be the types of plug-ins we have, screen format, fonts, language, whether cookies are accepted, and if certain features are turned on or off.

'We leak lots of information because when these systems were designed people weren't thinking about privacy,' Laurie continues. Instead designers were thinking how they could capture as much data as possible to make useful tools. 'If you had twenty possible plug-ins and people either did or didn't have each one then that's a million combinations right there. Make it forty and that's a million million so the chances that anyone in the world would have the same set-up as you could be one in a thousand.'[4]

I wonder how Laurie can be a privacy advocate on the one hand and a full-time employee of Google on the other. He claims Google currently works hard to respect people's privacy and handle their data carefully. But who knows what the owners in twenty years' time will do with the data that might still be lying around from now? He says this is even more of a danger with Facebook, who 'don't give a damn, they're quite happy to publish your stuff'.

Is that really true? I ask.

'Isn't that obvious? They occasionally make statements along the lines of "We respect users' privacy" but their actions prove that they do not. I think it's very clear that they don't give a damn. Facebook's business model is to know and reveal as much as possible about you because that's how it makes money and that's how it spreads. But the same argument applies to Google and search [engines]: Facebook have to be privacy non-respecting because that's how they succeed. If they were more privacy-respecting, then the next guy who was less privacy-respecting would take over from them. I do blame Facebook because clearly it's a conscious decision you make, but it's also inherently obvious that the most successful social network will be the least privacy-respecting. I mean, how do you find people? You find people by them blathering about themselves without any compunction.'

Finally, I ask Ben if he's actually on Facebook. This is where many privacy advocates come a cropper, I've noticed.

'I shut down my Facebook account after the last round of

4. If you want to know how unique your computer is then I suggest trying the EFF's Panopticlick at panopticlick.eff.org.

outrages. When they started tracking everything that you did, and everywhere there was a 'Like' button, I said that's it, so I closed my account and I block all Facebook cookies. But I talk to my younger, security-aware friends about why they still have Facebook accounts despite the fact they know all these abuses are going on, and the response is, "If I did not have a Facebook account I would never go to a party again," and apparently they care enough about going to parties that they are prepared to tolerate Facebook.'

After I left Laurie, I couldn't look at my computer the same way again. I stopped pressing all those 'Like' buttons for a start, knowing that Facebook was harvesting all that data for purposes unknown. I didn't want to get paranoid but it was clear there was a lot going on behind the scenes. Most importantly, what exactly were Google and Facebook up to? What were they doing with all the personal information they were harvesting on the world's citizens? There was only one way to find out.

Goin' to California

The Googleplex comes at the end of a short drive down Shoreline Boulevard, which features low-rise office buildings tenanted with tech companies such as Siemens and Microsoft but also smaller start-ups and a Starbucks amidst rows of Californian redwoods and oaks. Funnily enough, the week I arrived at the Googleplex in Mountain View, California, there's a poster on the noticeboard alerting employees that this week's Google film club offering is *The Lives of Others*, about an East German playwright and his girlfriend who are spied upon by the Stasi. I wonder if the irony is lost on those working here, but to be honest, if Google is the digital age equivalent of the Stasi, it's hard to be too worried. This is a place where grown men ride kids' bikes painted primary colours and leave them outside resting only on a kickstand; where there's an 'organic ecosystem garden' in the campus courtyard offering its harvest to everyone (in moderation) with a small sign, 'Don't be evil – leave some for others'; where there's a sand volleyball court outside and free food inside offering everything from hand-rolled sushi to Indian curry. The parking garage features pull-down plugs for electrical cars, though there's none in existence on the days I visit; instead a big blue Ford Mustang rumbles into the last remaining eco-car space.

It seems a benign, almost Disneyfied tech company. Hardly the evil empire of Total Information Awareness. Of course, this is Google's image and it is necessary if people are to feel comfortable using the website and sharing their personal information with the company. Google's public relations people appear open and friendly. Certainly that's the case with Anthony House, an American who studied modern history at Oxford University, and also Peter Barron, who heads up the European PR arm of Google and used to be head honcho at the BBC's current affairs television show *Newsnight*. The buildings I visit are all open-plan with only a few side offices, and walking around I spot CEO Eric Schmidt twice, ambling about, once outside and another time giving a tour to some foreign dignitaries. But it's not totally open. At reception there's a screen where visitors type in their name and who they're here to see. Once you've done that there's a screen to accept a non-disclosure agreement. I decide to decline. I can still enter the premises but my printed name tag is watermarked 'No NDA' and Anthony House says apologetically that this limits where I can go in the complex and that I won't be able to attend certain meetings.

NDAs are common in Silicon Valley and I wonder if anyone

has ever compared these NDAs with the companies' own privacy policies to see if there's a double standard between the importance they place on their own privacy compared to that of their users. The tech people here came to know sooner than most how easily information can be shared and spread, and their solution has been to protect a culture of internal openness by having privacy built in, through legal agreements governing disclosure.

During a break between my days at Google, I track down hacker and computer security expert Moxie Marlinspike. He's invented several programs to preserve an individual's privacy on the Net and phone but his particular bugbear is Google, and this led him to create Googlesharing, an anonymising proxy service that allows users to hide from Google by temporarily routing their traffic through other computers. Moxie is in his twenties, describes himself as 'one part sailor, one part hacker, one part pyrotechnician', and runs a company called Whisper Systems, creating privacy software for phones and computers. He's written Text Secure, an application for mobile phones that encrypts messages, another that encrypts voice calls and yet another to encrypt the entire phone's hard disk. We meet at the San Francisco hackerspace Noisebridge.

One of the reasons I've been given for people's seeming indifference to privacy online is that they don't see its lack causing any tangible harm. If you're in Iran or China or an autocratic state, then maybe you'd be worried. But in the United States, UK, Western Europe, Australia, what is there to worry about? I ask Marlinspike, what's so bad about Google?

'I do think Google are just trying to sell advertising,' he replies. 'There's nothing necessarily wrong with that. I don't think they have pernicious intent. They are not the evil empire – but the problem is that the data they have is really interesting to people who maybe don't have the same intentions. What most people don't realise is when you use Google or Facebook, you're not the customer – you're the product. *You* are what they are selling.'

And who are they selling 'you' to?

'With Google you never know. They're selling profiling information about people to advertisers, but maybe to other people as well.'

What sort of information?

'First of all they have all your search terms. If you don't have a Google account, they tag you with a cookie (a piece of text stored on your computer for identification), so if you're using the same browser they can correlate all of your history. They can determine who you are even if you're moving across computers – they have enough information based on your search habits as you're revealing a little about yourself every time you type in search terms. They use that information to build a profile of the people who are visiting their services.'

But they're doing all this to sell advertising, I counter. They're not John Poindexter. Surely no one's going to jail based on the info Google collects to sell you an ad?

'But people can go to jail, or worse. Because what ends up happening is Google collects more and more data, and data becomes attractive to all kinds of eavesdroppers. For instance, we saw this when Google got hacked by Chinese agents;[5] they were after that data, that's what they wanted.

---

5. Google was the target in December 2009 of a 'highly sophisticated' and coordinated cyber attack against its corporate network by hackers originating in China. The company claimed they stole intellectual property and sought access to the Gmail accounts of human rights activists. A US diplomatic cable published by WikiLeaks in December 2010 stated the hack was orchestrated by a senior member of the Chinese communist Politburo who didn't like what he saw when he Googled himself.

He puts forward the hacker mantra that attackers always win eventually because it's much harder to defend a system than attack it. Someone defending a system has numerous problems to deal with; a hacker just needs to find one vulnerability. The only thing you can do, as a defender, is to raise the cost for an attacker in terms of time, resources and money.

'What happens when you have extremely dense concentration of value in one place, like all this information in Google, it's hard to raise the cost high enough to prevent an attacker from going after it and because it's worth so much they are willing to invest a lot to get it. Chinese agents got jobs at Google offices in China posing as programmers or whatever – they compromised the whole office. That sort of thing is going to happen again and again. It's not just going to be eavesdroppers with a legal backing like the US government (who we also know from that attack are running lawful intercept stuff on their networks), but also people who don't have legal backing who are going to be attracted to this information.'

He tells me it's not just search terms that Google collects either. When we use Google maps and tie it to the GPS on our phones that data goes to the company. With Google Latitude our phones can report our real-time coordinates so Google knows where we are and our travel patterns. Google Analytics allows website operators to collect statistics on visitors, but the statistics are managed by Google. 'Every time you visit a website, a little connection goes off to Google that says, "I just visited this website, and by the way here's my cookie." Now Google knows not only what you're searching for when you use Google, but also what websites you visit even when you're randomly browsing the Web having nothing to do with Google services.'

I wonder why we don't hear more about this. It seems either

Google's press people are doing an excellent job or people just aren't that bothered. Moxie believes people do care about privacy but what Google has done is redefine the terms of the debate, by conflating privacy with anonymity. Google now issues press releases assuring users that all data is anonymised after nine months, but its definition of 'anonymise' is somewhat dubious, as it merely involves dropping the last octet of your IP address. This means if your IP address is 123.456.789.012, Google drops the 012. Moxie says it's 'totally easy' to correlate the other factors and de-anonymise the data.

He mentions Google's line about putting 'privacy in your control' by customising your account's privacy settings, but says the interesting thing is that in order to participate you first have to have a Google account – and as soon as you sign out, any restrictions you've set no longer apply. So if you care about privacy you're incentivised to stay logged into Google all the time. Also when you delete a Google account they don't throw away your information.

'They never delete anything, there's no value in deleting something. If they have information, it could be worth something somehow or some time and it's not going to be worth more if it doesn't exist. Storage is free, they've mastered this problem of storing massive amounts of data, they know how to deal with it, and it's not a burden to them at all. If you think about any data that you generate, who you are, what your preferences are, what your political leanings are, your health, your love life; you never know what's going to happen in the future with the people who control that data. Right now Google has this benevolent image – but that could change. The founders could retire and they could have a whole new regime there that decides they want to do it differently.'

When I return to Google, I put these points to Anthony

House, whose official title is Privacy Communications and Policy Manager for Europe, the Middle East and Africa. He confirms that Google's anonymisation policy is to delete the last octet but says removing this makes linking an individual log line to an individual computer impossible. They anonymise the IP address after nine months and the cookie ID information after eighteen months. 'We keep the anonymised logs for ever,' he admits, but denies that Google keeps everything for ever. 'When someone deletes an account, we wait a short period before deleting the account permanently to make sure the deletion request wasn't a mistake or a malicious act by someone who had gained unauthorised access to the account. We then delete the information permanently. The process is complete within sixty days.'

It sounds reasonable but it's not quite the privacy-friendly mantra I heard when I first arrived. I've had to talk to a specialist hacker to get this far, but at least Google is willing to talk, which is more than can be said for Apple.[6] Later I tell Moxie how Google responded and he says that it's 'absolutely false' to say that dropping the last octet of an IP address makes linking an individual log line to an individual computer impossible. 'This is a great example of their PR strategy to control this debate by defining the terms. If you ask any security

6. I'd asked Kevin Bankston, the EFF's senior lawyer, why Apple is so often left out of the privacy debates. He admits they've not done a great deal of work in relation to Apple. 'That's mostly a reflection of the fact that their practices are rather opaque and are not reported on often, while Google and Facebook, for example, while we don't always like what they do, they are at least fairly transparent about what they do or when they make changes.' In April 2011 there were howls of outrage when it was discovered that iPhones and iPads contained a program to harvest users' geodata even when they'd turned off permissions for geotracking. Apple claimed it was a bug and that it would be fixed.

expert whether this practice meets even the lowest possible bar or haziest definition for "anonymise", they'll tell you it doesn't, and yet Google persists to use this language.'

I ask him if he thinks Facebook is any better. He says they are similar, 'but at least Facebook doesn't try to pretend that they're looking out for our privacy'.

A few miles up Highway 101 and I come to Facebook's head-quarters at 1050 Page Mill Road in Palo Alto. It's not nearly as scenic as Google; instead it's a nondescript cement block. There are some 1,700 people working at Facebook, including 500 to 600 engineers and 20 to 25 lawyers, though not all work in this building: there's a smaller office a few blocks away. The entrance is dull – none of the fun and bright colours of Google – and at the front desk I'm met with two clipboards each featuring a dense non-disclosure agreement: one for meetings, the other for tours.

'We want to be an open place. A horizontal structure,' says Simon Axten, who works in privacy and public policy, as he leads us around the open-plan office, pointing out where Mark Zuckerberg sits in a group of desks (he's not there at the moment). Inside, the decor is very much the student dorm aesthetic. The conference rooms are themed by floor so on the ground they're all Internet memes (Om Nom Nom, Lolcats, Star Wars Kid), Vegas casinos on the first floor. There's a giant chessboard with toddler-sized pieces, bits of chain-link fence, a pirate flag, a poster of Bob Marley, some graffiti. One sentence reads, 'I drank the Kool-aid and it was good.' It's Rebellion Lite.

Axten's spiel about openness strikes me as hypocritical after I've just been asked to sign an incredibly wide-ranging and egregious non-disclosure agreement that basically defines as confidential anything and everything and for all time related to Facebook. I'm likely breaking it simply describing the inside of the building. It's almost an inverse of the privacy (or lack of it) the company grants to its users.

There's free laundry and free food, just like at Google, and Simon boasts how they stole Google's chef. There's a lot of poaching from Google. There's also a lot of celebrating – Facebook has passed its 500 millionth user the week I visit and it's champagne all round. The company is growing rapidly and moved into this building in May 2009.

Unlike Google, Facebook doesn't make public the number or type of requests it gets from governments to pull pages or hand over user data. Simon Axten won't say; neither did Richard Allan, Facebook's Director of Public Policy for the EU, when I'd met him earlier that month.

Facebook itself has changed its policy about privacy. Initially, founder and CEO Mark Zuckerberg was on record stating that user privacy was 'the vector around which Facebook operates' and that the website's popularity hinged on the fact that it promised people their information would only be visible to friends. The company stored customer data solely in its site, believing that this was needed to make users feel comfortable enough to share more information with a small number of trusted people.

In December 2009, after hitting 350 million users, the company radically changed policy, shifting the default privacy settings so that your name, profile picture, gender, current city, networks, Friends List and subscribed pages were publicly available so anyone on the Web could see them.

'People have really gotten comfortable not only sharing more information and different kinds, but more openly and with more people,' Zuckerberg told an audience in January

2010. 'That social norm is just something that has evolved over time.'[7]

It's true that society's thinking on privacy is changing in relation to the networked Web, but I'm not convinced Facebook is simply reflecting society's views or that we all want to share everything with everyone. While I'm wary of fetishising privacy, I also think it suits Facebook's corporate interests to have as much user data available as possible for targeted ad revenue generation. In case people don't know, Facebook already tracks all your posts, statuses and 'Likes' to select which ads appear on your pages and in March 2011 it began a test phase for mining real-time conversations (status updates, comments and wall postings) to target ads to users. This has created some concern that for a few cents, advertisers, or someone posing as an advertiser, can access customer data. Stanford computer science researcher Aleksandra Korolova detailed how she used Facebook's targeted ad service to discover if a female friend was gay (even though her orientation was set to 'friends only'.) Facebook shares with advertisers publicly accessible data like age, location, education and interests so Korolova was able to create a campaign that was targeted to women interested in women. Her friend's name came up on the list of people who were shown the ad.

When I left California I certainly had a better idea of how Google, Facebook and others like them are storing and using our data. There's a lot more to know but while these companies are keen to harvest our information, they are less willing to provide their own. Pressure is growing on them to be more publicly accountable but what about the smaller companies?

7. http://tinyurl.com/yadgm8x/

There is an entire industry that exists in relative obscurity trading in what is becoming a very lucrative commodity.

Data brokering: your personal information is big business

In London I witness one such potential trade. It's at a restaurant in St James's, an upscale place with a painting at reception of a woman lying naked but for a white silk blindfold over her eyes, her body arched in anticipated sexual pleasure. It's an area of the city known for its posh private gentlemen's clubs, cigar shops and tweed menswear catering to the landed gentry. It may be a vestige of old aristocratic England – and a world away from Google's colourful California headquarters – but at my table I am privy to a very modern conversation over lunch: the buying and selling of personal data.

'I've got a friend who could probably get it to you for seventy-five or a hundred,' says a dapper man in a cufflinked white shirt. He means £75,000–100,000 and he's talking about a copy of Britain's full electoral register in digital form.

The other two people at the table aren't biting.

'Thinking about it I could probably get it for fifty, path-validated,' he offers.

The others decline and this particular transaction isn't successful – but others are (often for much more money and much more data). Yet the general public has very little know-ledge of and even less control over this industry, which derives its income from our personal data.

In the UK the full electoral register is a public document in name only. Only registered political parties and campaign groups authorised by the government, and credit reference agencies such as Experian, can obtain digital copies of the entire database. This arrangement is meant to strike the right balance between privacy and accountability but it has not stopped data brokering.

While the British may not like their details being shared with companies or the general public, they tend to be less concerned with details harvested and shared by the state. This is partly due, I believe, to bureaucrats' clever repositioning after the Second World War from enemy of the people to friend, with the roll-out of universal welfare systems such as the National Health Service.[8] Even though the British are the most surveilled in the world, the population has only recently mobilised against CCTV and state databases as an invasion of their privacy. The practice of building immense databases of citizen data to solve every social need reached its zenith with the disastrous NHS Spine, an ill-thought-out plan to create a central electronic database of every person's medical record with costs estimated at between £12.7 and £20 billion. A backlash against these huge and hugely intrusive state databases took shape during the May 2010 general election and the new government abandoned several of the less popular databases, including a national identity register.

The US government was, until recently, stymied by the Constitution's protections against unwarranted state interference. Through a combination of measures such as the Patriot Act, subpoenas, foreign intelligence sharing and the purchase of data from private companies, American officials are now matching, if not exceeding, other countries' penchants for surveilling their citizens. Dozens of commercial data brokers compete to sell personal information to the government and

8. See Chapter 1 of *The Silent State* for a discussion of the way the state redefined its role after the war.

anyone else with the money to buy it. And unlike in the UK, American public records such as electoral rolls and drivers' data are bought and sold freely on the open market.

Brian Alseth knows a lot about data mining. He's a lawyer and self-confessed geek who worked on the Obama campaign's data team – a campaign that became famous for its skilled use of technology and social networking to drum up votes and cash. Alseth now works as the technology and liberty director for the Washington state branch of the American Civil Liberties Union and I visited him in his downtown Seattle office to discuss the data mining industry.

The first thing he shows me is a 406-page 'data dictionary' from Acxiom, one of the biggest data brokers in the US and the world. The dictionary outlines some 10,000 elements of personal data that can be bought entirely legally – think of each as the column header of an enormous database. Number 7724, for example, is current affairs/politics. This shows every person in a household who has an interest in current affairs. As I look through the pages there are hundreds of different categories, from individual income and the value of your house to religious beliefs and even whether you prefer diet soda or full fat. Each data element lists where it came from, be it public records, surveys, registrations, telephone directories, business records, government transactions, etc. Some companies also do data cleaning, which involves ringing people to check accuracy.

Data brokers are a bit like used car salesmen but instead of offering this vehicle or that, they offer all sorts of personal data. You want a list of white people looking to buy a car next year? Here you go. Old Jewish ladies who like cats? Easy. Gay Republicans with incomes above $100,000? You name it, they've got it. And if they don't have it they model it, by putting several data elements together using an algorithm to come up with a predictive model of behaviour. It's the same technique used by Amazon to predict what books you might buy in future based on what you and others like you have bought in the past.

'That's the goal with any of this – political data or consumer data – to predict the future,' Alseth says. 'You want to know, based on who someone is, how they will behave in the future or who else will behave like them. So with political data, what we would do is first create a sample size of what we call hard IDs [identities] and then build a model based on that.' He offers an example from his political days where a campaign would get a sampling of 5,000 hard IDs with their voting patterns and match this up with additional personal data bought from another company, then build a model.

'We look down the list and see out of those people 0.02 per cent of those who own cats are more likely to vote for our candidate, or we might see a 0.4 per cent increase in support among people who drink bourbon versus people who drink vodka. Or heavy fast food users, or people who buy things on the Internet. All this creates data points and once you figure out a model you can then multiply across the table until you have one standardised score as to what a likely support model is. Once you have that algorithm based on the 10,000 data points you can turn that map around to everyone not talked to and apply the same thing and extrapolate out to the other 5 million people in the state. As you keep moving through the campaign and IDing more people and finding out more about people, the model keeps adjusting and getting better.'

The first places most data brokers begin when building a database are telephone books and public records. In the US, every state has a voter file of every registered voter and by law these must be publicly available to anyone to be used for any political purpose. Campaigns pick up these electoral registers and then go out into the field asking people about specific issues: this person is interested in gun rights, or reproductive rights, or gay rights. All of this creates additional data points and then the campaign buys in additional information from major brokers like InfoUSA or ChoicePoint. InfoUSA also obtains data from First Data Resources, which processes the majority of the world's credit card transactions, so they also have reams of information on what people buy and how much they spend. Data from companies such as Catalist and VoteBuilder is then bought to check the data and get the files to match.

Data in politics is nothing new – Alseth's mom told him stories about campaigning in the Kennedy election in the sixties where coloured notecards were used to record voters' preferences: red cards for people opposed, white cards for those you still had to talk to, and blue/green cards for supporters. What is different today is the volume of data that can be collected, the ease of sharing and the sophisticated analysis – all of which are a result of digitisation. Companies can sell all this, instantly, to you, me or anyone willing to stump up enough cash. Some restrictions do exist – certain states have limits on what you can do with driver's licence data, for example, but most don't. Social security numbers typically aren't added to commercially available data sets but the companies have them even if they are not selling them. Data brokering is a business, the commodity is information, and in terms of what can and can't be sold, Alseth describes the US as 'a feudal wasteland'.

'I was chastised once when I asked one of these data shops if they had privacy compliance officers, and they said, "Now Brian, I would never call you out but why ask a question you already know the answer to? We had enough trouble getting any sort of quality assurance in here." It's just not something they worry about. For the most part people aren't aware that this much data about you is out there already, based on just simply living your life.'

As a relatively new industry, data brokerage has been quick to take advantage of the changes wrought by digitising data. While much of this data wasn't exactly secret, most people probably assumed that their credit card transactions, for example, were 'private' between them and their credit card company or that a public record was limited to a specific geographic area. Now all this data which may have once been considered private due to physical constraints is globally available. Occasionally, the Federal Trade Commission will wade into this unregulated world to rule on specific acts by some companies but there are, as yet, no federal laws. Amber Yoo, a spokeswoman from the consumer protection organisation Privacy Rights Clearinghouse, has described the data broker industry as 'the Wild West', such is the lack of legislation regulating what they can and can't do. Each data broker 'has their own mechanism for opting out', Yoo continues, 'so you have to go company by company. Some of them don't provide any mechanism to opt out, making it impossible to entirely restrict all of your information. Even worse, some make you pay to have your own information removed.'[9]

Along with Acxiom, ChoicePoint and InfoUSA we can also add Reed Elsevier (a UK and Dutch registered company that is the parent of LexisNexis, Westlaw and ChoicePoint), Experian, Dun and Bradstreet and CACI as some of the world's largest data brokers. Acxiom developed its own grid-computing

9. http://tinyurl.com/6h2a3q6

technology to help it manage petabytes (1 quadrillion bytes, or 1,000 terabytes) of data. It also sells data to brokers such as USAData, through whose portal anyone with a credit card can buy marketing lists of consumers according to geography, demographics and interests.

Although the majority of this data is used for targeted advertising, a big customer is the US government. The Internal Revenue Service and departments of Homeland Security, Justice and State paid $30 million in 2005 to data brokers, most for law enforcement and 22 per cent for counterterrorism. That same year the IRS signed a five-year, $200 million contract to tap into ChoicePoint's databases to locate assets of delinquent taxpayers.[10]

There are of course legitimate public service reasons to collect and model data, for example figuring out where services are required or how tax changes will affect particular communities. Where this stockpiling causes concern for groups such as the ACLU, though, is its use to label individuals in a way that impinges on their lives and against which they are largely powerless to defend themselves. Companies like IBM are investing billions into their analytics department and marketing products to law enforcement agencies and governments which Alseth claims resemble 'pre-crime' from *Minority Report*. This is the biggest danger of governments' use of personal data – as the grounds for interference in people's lives based not on what they have done but on what they might do in future, abolishing Enlightenment ideas of free choice, probable cause and innocent until proven guilty.

'Scores are being assigned to us that we don't know, and that is guiding how we are treated,' warns Alseth. 'Whether you

10. http://tinyurl.com/5tegjwj

end up on a watch list, whether they stick a GPS tracking device under your car without a warrant, all of these things. Whether or not you get an offer in the mail for your favourite soda.'

And here's the rub – we have no idea how far along the US government's use of this modelling is because they won't tell us. Alseth says, 'They are further along than we would ever imagine, especially the National Security Agency.'

Privacy law in Europe is, in principle, more robust than in the US, with all EU member states implementing the European Data Protection Directive. This purports to give people the right to know what data is collected on them, by whom, for what purpose and who it is shared with; and should enable them to access their own personal data and correct it.[11] However, in practice, these rights have been frequently subverted so that states and large corporations continue to stockpile personal data on people without accountability. There are many exemptions to deny the private citizen a right to his or her personal data. However, when private citizens try to hold public officials to account on, say, how they spend public money, these officials invariably refuse to provide the information, claiming to do so is an invasion of their privacy. Indeed, this was the reason cited by Members of Parliament when they refused freedom of information requests for details of their publicly funded expenses. Additionally, the European Court of Justice ruled in November 2010 that the identities of farmers in receipt of €55 billion a year in farm subsidies (more than 40 per cent of the EU's entire budget and about €100 a year for each EU citizen) was a violation of their privacy. Never

11. In March 2011 the EU also announced that it was seeking to enshrine a 'right to be forgotten' in law, whereby social networking sites would be obligated to delete personal data at users' request.

mind that it was public money and the public had a right to know who they'd given it to. There is another disadvantage of the European system: precisely because there is a law, the data industry is less forthcoming about how it operates, hence the lunchtime meeting to discuss a database that one could easily buy on the open American market. If Americans don't like this they can lobby to create a privacy law. In Europe it's difficult to lobby because we don't know the full extent of this shadowy trade.

In the digital age, it is true that privacy no longer exists – at least if we take privacy to mean that no one knows anything about us. Someone, somewhere knows something about us or can figure it out, especially if we've ever been online. It is becoming increasingly difficult to live off the grid. We need to be aware that everything we do online can be stockpiled and sold and we have very little idea to whom or for what purpose. The law is a long way from catching up with the changes in technology; even in Europe, with its Data Protection directive, citizens have little power to uncover where their data is going.

The Brandeis definition of privacy – the right to be left alone – is most in jeopardy with the advent of digital databases. Our jobs, financial future, movement and freedoms are increasingly being determined by how we score on vast data sets that are traded and to which we have limited access and not much public oversight.

The tangible harm is not always apparent. It's easier to look at the rise of online tracking and data harvesting not as a battle between security and privacy but rather convenience versus privacy. Many people like logging into Amazon or a movie site like Netflix and getting recommendations of what they should read or watch. But there are consequences that will become more apparent. Mostly these will revolve around a growing inability to be simply *left alone*.

Apart from changing the law, another solution is to build privacy into products, as suggested by Ben Laurie, John Gilmore, Moxie Marlinspike and nearly all tech people who share an interest in privacy. You wouldn't want a house built without locks just because burglary is illegal, so why use a piece of software without inbuilt data protection? Companies and governments are at pains to tell us that they value our personal data, that it will be kept confidential and secure, that they won't share it with anybody (except perhaps a few 'trusted' third-party business partners or law enforcement), but guess what – they've been hacked. Or an official has lost his laptop. All the worthy statements in the world can't stop the data being sold or shared around the globe then.

What can we do? In Europe you can make a request under the Data Protection Directive to any organisation for a copy of your personal data, though from my own experience this is not always fruitful as enforcement of the law is almost non-existent. You can also avoid giving out personal information to survey takers or online. If you're curious about which Disney princess you are on a Facebook app – do they really need to know your date of birth? Do they need to know what magazines you subscribe to? If you're surfing the Web, turn on Adblock. Use the tools that are available to you to protect the private information that is important to you.

Mostly it's about paying attention. In the digital age information can flow freely, and it's up to us to decide how and where to limit this flow. Your data is your digital DNA and just as we wouldn't like our biological DNA shared with the world, we're probably not keen to have it bartered openly by a US company or traded covertly at a St James's lunch. Unless we pay attention we remain like the naked lady in the painting – transparent to all but unaware who is watching until we take the blindfold from our eyes.

After meeting Julian Assange, Nick Davies returned to Britain and the *Guardian*'s headquarters at King's Place where he talked to *Guardian* editor Alan Rusbridger about how to set in motion the paper's entry into the world of large-scale digital data journalism. Rusbridger was keen to get the *New York Times* on board for multi-jurisdictional protection as the *Guardian* had already been slapped with an injunction by Barclays Bank when it tried to write about their tax evasion scheme.

'I rang Bill,' Rusbridger told me, referring to the *New York Times* editor Bill Keller. 'None of us knew about encrypted technology so I texted him and asked, "What's the most secure way of speaking?" He didn't know either. We ended up just talking on the landline. I said to him, "We have an amazing story we want to work on with you but I don't want to talk about it over the phone. It's about Iraq and Afghanistan. Will you send someone over?"' Eric Schmitt, the *New York Times*' terrorism correspondent, was on the next plane. Initially there was some scepticism about working with the *New York Times*, a paper with a reputation for both courage and arrogance, but as David Leigh later put it, 'The honour of the top guys was involved so it was all going to be OK.' The German magazine *Der Spiegel* would also be part of the deal, at Julian's insistence.

Nick was full of energy and enthusiasm for the project: 'We need an office. We need to call the correspondents back from Afghanistan,' he told Rusbridger. David – still miffed that Julian hadn't agreed to work with the *Guardian* on the Apache helicopter video – was less optimistic. ('It will all end in tears . . .') Alan Rusbridger was simply intrigued.

Nick and David kept asking Julian Assange to come to London so they could talk about the data and the technical issues but Julian put them off, telling Nick by way of explanation that he'd impregnated a woman in Paris and was going to return for the birth of his child. For whatever reason there was no sign of him until 7 July when, after I failed to provide the necessary hospitality, he headed to the *Guardian* office.

'He did this Peter Pan thing,' David recalls. 'He pitched up with his great big rucksack with his desktop computer in it and no plans and nowhere to stay. He ended up in my spare bedroom for a few days. Then I put him into a hotel in Bloomsbury.'

In the meantime, Nick and David had taken charge of turning the data into articles. It was a laborious process and the two veteran journalists had a baptism by fire with computer-assisted reporting.

'I was appalled when I first saw these spreadsheets,' David told me later. 'We couldn't even get to first base because we couldn't download it. Then we thought we'd downloaded it as an Excel file and it turned out we didn't know that we only had Excel 2004, which only reads 65,000 rows. So we thought that was it. The Germans were then talking about files we didn't have and we discovered there were 90,000 files. Anyway at that point we stopped pretending we could understand any of this and called in the guys from our IT department.'

Harold Frayman, a long-time subeditor at the *Guardian* and computer expert, built a database so the half-dozen reporters could search the Afghan war field reports by date and other identifying factors. The database contained 92,201 rows of data, meaning reporters couldn't simply read through the information; instead they had to find new ways of analysing it.

'We learned how to filter columns and all this basic Excel stuff none of us knew before,' David said. 'Eventually we got the data to speak to us. That's quite exciting when you find ways of interrogating it and it will answer. It's like going to Mars and finding a way to have a conversation with a Martian.'

'All kinds of blood-curdling warnings were given to me about what we could be getting into,' Rusbridger told me. 'At the time our main concern was the American Espionage Act and the British Official Secrets Act.[12] We had a feeling we'd be injuncted.'

Deputy editor Ian Katz says the news team worked on the assumption that the intelligence services of one of the countries 'knew what we were up to. But it became clear they had absolutely no idea.' The journalists working on the Afghan logs didn't use email or phones; instead they stuck to Skype, believing it was less likely to be intercepted. So pervasive was the fear of a pre-publication injunction that all the stories the team uncovered, each of which could have made a splash on its own, were put into a single day's newspaper. It was information overload and diminished from the overall impact of each story, but Rusbridger argued that the paper had to publish in one gigantic dump for three reasons: 1) Julian Assange was going to publish all the raw documents, 2) the German weekly *Der Spiegel* was publishing everything in one go, and 3) 'after Barclays [Bank] the [*Guardian*] lawyers were saying, "They will come in and they'll seize all the notebooks and computer material." They can do that in Britain.'

So at 10 p.m. on 25 July, the *Guardian* posted all of its Afghanistan stories online, including revelations about the

12. Both laws were passed around the time of the First World War: 1917 for the US Espionage Act and 1911 for the Official Secrets Act.

secret American special forces unit that exists to carry out 'kill or capture' operations; that the coalition was increasingly using Reaper drones to hunt and kill targets from a remote control base in Nevada; and that the Taliban had acquired deadly surface-to-air missiles and were causing increased carnage with a massive escalation of their roadside bombing campaign, killing more than 2,000 civilians – all of which the US Army had kept quiet.

Later that night, WikiLeaks published 76,000 of the raw reports on its website, holding back at the last minute 15,000 of the sensitive threat reports after sustained pressure from the media partners.

The next morning Julian Assange heads out from the mews house where he is now staying, slips round the corner and walks up the narrow stairs to the top floor of the Frontline Club, where he's organised a press conference.

'Holy fuck,' he mutters with a satisfied smile as he walks through a scrum of journalists. I'm in the audience and the difference between this and Norway is stark. There's not an empty seat to be had among the sixty or so reporters, photographers and TV cameramen who have their bulky TV cameras propped along a bank of tripods at the back of the room. Assange pulls back his shoulders, stands straight and adopts an intensely serious expression to match the content of what he is here to discuss.

'Some of the morning papers,' Assange says as he stands at the podium. Behind him is Don McCullin's famous black-and-white photograph of a shell-shocked Vietnam veteran. Assange looks out at the assembled journalists. 'Fourteen pages in the *Guardian*. Seventeen pages in *Der Spiegel*. The *New York Times*, the *Telegraph* also on the front page.'

A flurry of flashes as he holds up the *Guardian* newspaper and his white hair shines.

He's asked what is the biggest revelation. It's not one single incident, one single mass killing. 'That is not the real story of this material. It's war. It's one damn thing after another. It is the continuous small events, the continuous deaths.'

One of the most noteworthy stories found by Nick Davies and the *Spiegel* reporters is the existence of the special forces 'kill or capture' squad Task Force 373, which is essentially a group of agents tasked with implementing a US assassination list in Afghanistan. Not only do the documents reveal the existence of this extra-judicial squad, they also shed some light on how people wind up on such a list – through the recommendations of regional governors or intelligence authorities based on little evidence and no judicial review.

Another reporter asks if he's saying these are war crimes. Nick Davies has gone on record saying he believes Task Force 373 committed war crimes, but here Assange is guarded. He ignores the question; the reporter shouts out again, asking why he won't answer.

'I don't find that a legitimate question,' Assange responds.

'Hang on, you're here talking about freedom of expression—'

'Next!' Assange interrupts.

'—and you're not answering. That's a little hypocritical.'

'You have to phrase your question—'

'You used the term "war crimes" and now you're refusing to explain why.'

'You have to wait your turn. And phrase your question in a meaningful way.' Finally, after a third attempt, the reporter gets an answer: it is up to a court to determine if the actions are crimes but 'prima facie' it looks as though at least one incident involving Task Force 373 is a war crime.

Next comes an easier question: what is the motivation for the source in leaking this classified material? Assange responds that his source wanted to bring these matters to the world's attention though he says he doesn't know who the source is. (He also claims that he has committed funds to Bradley Manning's lawyer, though this is not true.[13])

Describing the documents, Assange mentions that 'between us we've read 1,000 of these reports properly', which instantly prompts someone to ask if publishing 76,000 reports he hasn't read is a responsible thing to do.

'We have a harm minimisation policy,' Assange says. 'Our goal is just reform. The method is transparency. We don't put the method before the goal. We do not do things in an ad hoc way.'

This sounds reasonable, and it is on this basis that I go on BBC's *Newsnight* later that week to defend WikiLeaks' publication of the material. However, when I later ask John Goetz from *Der Spiegel*, David Leigh and Nick Davies from the *Guardian*, Birgitta Jónsdóttir, Daniel Domscheit-Berg and a number of other WikiLeaks people about this 'harm-minimisation policy', ad hoc appeared to be exactly how it was done.

As I stand at the back listening to all this I feel a strange mix of admiration and unease. I'm struck by the similarity of Assange with the politicians he claims to loathe, picking and choosing which questions he'll answer while dismissing others as 'illegitimate'. I'm impressed by his bravery and remarkable

---

13. A donation of $15,000 was not actually paid to Manning's legal defence team until January 2011, after a direct appeal by David House.

fearlessness in publishing such material in the face of constant threats, but the self-righteous zeal is becoming disturbing. I believe it is possible to reject authoritarianism and a culture of 'just following orders', to think for one's self about what is right and wrong, without becoming a monster or a megalo-maniac. But if thinking independently means believing yourself to be above the law, it becomes a trickier area to navigate.

8

The Information War Begins

Berlin, October

I walk beside a canal to a covered metal bridge opening up like a steel throat from which the sad song of an accordion sings; on the other side the player sits on a crate in the shadow of an elevated railway. Down a ramp to a cobbled path, past graffitied walls and into the clean lines of neoclassical apartment buildings three and four storeys tall. At 11 Marienstraße a white stone building is etched with a design that looks rather like a circuit board and to the left of the dark brown door there's a sign – a blue square oozing knotted wires. Next to the door there's a bell-push and I press the one marked 'Chaos Computer Club'.

The CCC is Europe's largest hacker organisation and also one of the oldest worldwide, set up in 1981 by hacker Wau Holland and others who predicted the importance digital technology would have in people's lives. The CCC is famous for publishing the security flaws of major technologies, from chip and PIN to smartphones. You want to know how to set up your own GSM transponder – here's the place to learn (within legal constraints, obviously). Among some of their more noteworthy 'hacks' is pulling the fingerprints of the German Interior Minister from a water glass and putting them on a transparent film that could be used to fool fingerprint readers. The biometric system proved more fallible than the ministry claimed. The club also worked with activists for voting transparency (including Rop Gonggrijp) to expose flaws in computerised voting machines. These were later ruled unconstitutional in Germany and abolished in Holland. The CCC isn't just about technical hacking, it's a hub of political activism based around a few common goals: transparency of government information, individual privacy and the removal of information-sharing restrictions such as the all-rights-reserved copyright model. Many of these hacks are demonstrated at the annual conference at the Berlin Congress Centre where several thousand hackers converge just after Christmas, and it was here in 2007 that Daniel Domscheit-Berg first met Julian Assange in person.

I walk through a courtyard surrounded by apartment blocks and into the CCC. There's a collection of desks pushed together to make a large table with coiled wires, surrounded by half a dozen well-worn office chairs and sofas. The walls are exposed brick with sections painted white and it feels airy with 10-foot ceilings. The L-shaped room is home to mostly young men, though there are a few older hands who drop in.

This is the physical space for a much larger online community but it provides the core, a homely hub for hanging out and even crashing overnight on the many sofas. The floors are rough wood and wires snake everywhere. A CCTV camera hangs from the underside of a steel girder; speakers are strapped to another. The lighting is a collection of paper lanterns and industrial spotlights. The popular hacker's drink, Club-Mate, is piled in crates at the back and dispensed from a defaced Coke machine. Someone has drawn a tic-tac-toe board in the dusty screen of an old video game, Ideal Twinline, and underneath, the words, 'How about a nice game of chess?' There are posters on the walls: 'Liberty waits on your fingers', 'Keep on blogging', and a stop sign with the words, 'Stop RFID'.

I've come to meet Daniel Domscheit-Berg, the former partner of WikiLeaks who's just been 'suspended' by Julian Assange. One of the topics we discuss this cool autumn afternoon is who now 'owns' WikiLeaks. Assange has changed his tune about freedom of information and leaks now that he's become a gatekeeper. He has told his former colleagues that he 'owns' the organisation, including all systems and content. This is contested by other WikiLeakers, notably a person who wishes to remain anonymous and to whom Daniel refers as 'the Architect', one of the technical team who built the site's secure submissions platform (among other things) and who subsequently decided to take it back, citing Assange's irresponsible behaviour. The Architect and Daniel are also in possession of some of the backlog of leaks that built up while Assange was focusing on a handful of 'mega-leaks' about the US government.

The question of who owns WikiLeaks has stymied lawyers from organisations as diverse as Julius Baer Bank and Trafigura to the Church of Scientology, who all sought fruitlessly for a registered legal entity they could sue. That had always been the great strength of WikiLeaks. As the organisation disintegrated, it was becoming a weakness.

'We are a stateless organisation,' Daniel tells me. 'There isn't one entity that owns everything. That's the whole point. Who has ownership? That is an open question. I don't think Julian has access to the original site. It's a technical matter and it was designed so that technical is split from content. When Julian made clear to us he controls the domain name that didn't go down well at all.'

Daniel is referring to WikiLeaks' formal registration of the WikiLeaks.org domain name. The original site registrant was not actually listed as Julian Assange, but a man called John Shipton who was later revealed through Australian court documents to be Julian's biological father. When Julius Baer Bank tried to sue WikiLeaks in February 2008 for publishing documents alleging tax evasion and money laundering by its Cayman Islands clients, Assange refused to respond to the litigation, claiming he was not the legal owner of WikiLeaks. Lawyers for the bank served court papers on the only 'owners' they could find: the ISP Dynadot and, through them, the registered domain name holder John Shipton. They also targeted Dan Mathews, a PhD student at Stanford University who was listed as the administrator of a Facebook group for WikiLeaks. Mathews is now an assistant professor of mathematics at Boston University and I went to visit him there to find out more about this case – the only time an organisation partially succeeded in getting WikiLeaks off the Internet and into court.

Mathews had been president of the Melbourne Mathematics and Statistics Society, a student organisation at the University of Melbourne, where he first met Julian. After graduating Mathews moved to the US to take a PhD at Stanford, but Julian kept in touch with him, and in December 2006 recruited him to work on the very first WikiLeak: an apparent secret plan approved by Sheikh Hassan Dahir Aweys, the rebel leader of the Islamic Courts Union in Somalia, to hire criminals to assassinate government officials (its authenticity has been questioned, with the suggestion it was a smear by US Intelligence, designed to discredit the Union, fracture Somali alliances and manipulate China). Mathews also worked on the publication of the Guantánamo Bay prison manuals and documents detailing corruption in Kenya.

He told me that in February 2008 he was sitting in his office at Stanford when a man appeared at his door with a stack of paper. 'I thought, this student has a lot of homework.' The man asked if he was Daniel Mathews. When he responded that yes, he was, he was served with what turned out to be a huge pile of court papers from a law firm trying to sue WikiLeaks. 'I thought, this is clearly a mistake . . . What does this have to do with me?'

He emailed the law firm saying there must be some misunderstanding. They wrote back within minutes stating that WikiLeaks listed him as an officer of the company on their Facebook page and as an officer they were entitled to serve him. Just in case he didn't get the message the first time round, they sent all the papers to him again electronically. Needless to say, he was more than a little concerned to be on the receiving end of a lawsuit from a Swiss bank.

Mathews got in touch with Assange, who responded by giving him the names of various First Amendment groups in the US who could support him. But apart from that he was on his own. Assange would not be representing WikiLeaks in court, so the bank was going after whoever it could, even if that meant the 'officer' of a Facebook group. Mathews was a defendant whether he liked it or not. He realised he'd better get a lawyer. He contacted the Electronic Frontier Foundation and the California First Amendment Coalition, along with many other organisations, and crossed his fingers.

In the meantime the bank's lawyers successfully applied for a broad injunction locking the WikiLeaks domain name and preventing its transfer, effectively removing it from the Web. The ISP Dynadot was a small business whose owners weren't interested in fighting a Swiss bank, so they agreed on their part to settle the lawsuit and the injunction was granted.

By now Julian Assange was putting out press releases about

WikiLeaks being pulled off the Web by a bullying Swiss bank, an injustice they would be fighting with a full legal team. While this strategy succeeded in portraying WikiLeaks as the victim, it also made it more difficult for Mathews.

'Now when I rang up the First Amendment Coalition they said, "WikiLeaks says it has twelve lawyers, why don't you just get one of them?"' But not a single one of Assange's small army of lawyers seemed to exist. 'I don't know what the situation was but certainly Julian was referring me to these people and they were referring me back to WikiLeaks, so from my point of view you could regard that as being screwed over by everyone involved.'

The case was scheduled for another hearing at the end of the month. On 29 February it looked as though Mathews was going to be the only defendant in court and he was more than a little unhappy to be carrying the can for WikiLeaks. Fortunately the media's portrayal of the judge's injunction as a flagrant abuse of law and a violation of the First Amendment led to, as Mathews describes it, the cavalry arriving two days before the hearing. Eleven groups, including the ACLU and EFF, joined the case as friends of the court, then the night before the hearing, lawyers representing John Shipton asked to attend as well. In court, however, the latter emphasised they were representing Shipton not WikiLeaks, as, 'All he does is own the domain name. He doesn't operate the website.'

The issue of who owned WikiLeaks had stumped the bank's lawyers and proved to be just as troublesome in court. Julius Baer's lawyers tried to argue that under ICANN[1] rules the person registering a domain name had ultimate liability for

1. The Internet Corporation for Assigned Names and Numbers is a not-for-profit public-benefit corporation set up in 1998 that coordinates the Internet's naming system.

the website at that address and as such was the legal owner. In the absence of that person (the bank's lawyers could not get hold of John Shipton) the liability was on the domain name registrar, the ISP Dynadot.

Dynadot's lawyer took issue with this claim, explaining that what an ISP does is merely register a domain name – which in this case was wikileaks.org – and allow that name to be translated into an Internet Protocol (IP) address, in this case, of a computer owned by WikiLeaks located in Stockholm, Sweden. 'In that sense,' argued their lawyer, 'I think Dynadot would be considered an access software provider because we do provide enabling tools that allow a computer user to transmit, receive, display, forward or translate content,' and as an access software provider they were not liable for third-party content.

Ultimately WikiLeaks was 'not a suable entity', explained another lawyer, this time from the Public Citizen Litigation Group, and if it was any kind of entity it was more likely an unincorporated association. In such cases the ownership of the organisation is based upon its membership. The bank's lawyer admitted as much, stating in court that the temporary restraining order against WikiLeaks had been nothing more than a legal strategy to force the website's legal owner into court. It was crafted, said the bank's lawyer, 'in the only way in which it could be enforced: through one of its third-party providers or agents to stop rendering the services or lock the account or strip the information until such time as WikiLeaks complies'.

This is worth noting because the same tactic would be used against WikiLeaks in December 2010 as the site began publishing the US diplomatic cables.

Holding the First Amendment hostage in order to accomplish

some other litigation was heavily criticised both in court and in the public arena. Julius Baer Bank came in for a pasting and were described in the press and on blog posts as the enemies of free speech, bully boys who were trying to shut down a tiny website. Hundreds of mirror sites were set up by outraged netizens who reposted the 'banned' material. A strategy designed to limit the information's exposure had done precisely the opposite in a way known as the 'Streisand effect'[2] whereby attempts to suppress information lead to even greater publicity. Even so, such injunctions are still used (in England in particular) and by their existence they reveal an ignorance of or refusal to deal with the realities of our networked digital age. The Internet allows information to be shared beyond national legal jurisdictions, so a court order forbidding publication in one national jurisdiction simply means those documents will pop up in another. It's like whack-a-mole where you might think you can hit one mole in one hole, but it only leads to more moles in more holes elsewhere.

The bank argued that the injunction should be permanent and the mirror sites punished for contempt of court, but the judge ruled against this as impractical and the injunction was dissolved. In his ruling he urged the parties involved to 'take a deep breath especially the plaintiff, and determine . . . whether there may be other ways to achieve the same goals. And maybe

2. The singer in 2003 unsuccessfully sued photographer Kenneth Adelman and Pictopia.com for $50 million, citing privacy violations over an aerial photograph of her Malibu mansion. The photo was one of 12,000 from a public collection that the photographer said he took to document coastal erosion. Outrage over the lawsuit generated so much publicity that the photo which had previously existed in some obscurity was now clicked on by millions of people all over the world.

there aren't. Maybe that's just the reality of the world that we live in, that when this genie gets out of the bottle, it's out for all purposes . . .'

The bank dropped its lawsuit against WikiLeaks on 5 March that year and Mathews was finally in the clear. 'Talk about a chilling effect,' he told me. 'I'd been drawn into this, and it had very little to do with me. I'm glad the court came to the right result but that was a little bit hot. It goes to show that within activism, you have to evaluate the consequences. A bad such consequence is suffering some sort of sanction because of actions over which you have no control.'

This puts into perspective Assange's claims of WikiLeaks ownership in the autumn of 2010. How could he invoke ownership laws which for years he'd claimed didn't apply to WikiLeaks? If the legal status was, as the lawyers argued in the Baer case, more like an unincorporated association, that meant the members were just as much legal owners as Assange.

Back in Berlin's CCC, Daniel Domscheit-Berg tells me he doesn't want to get into a 'bullshit argument' with Julian about stealing business secrets. He's not fighting for the money or anything else. The Architect is simply taking back what is his and the leaks are stored for safekeeping. The feedback Daniel's receiving is that people submitted leaked material to WikiLeaks, not to Julian Assange personally, and they did so based on what the site stood for, its security and technology. The technicians who built it therefore feel a responsibility for the material to be in safe hands.

WikiLeaks is disintegrating. Daniel has left and so have others. There are a number of reasons why the organisation is coming under strain but the common factor is the autocratic leadership of Julian Assange. His unilateral decision to publish the Afghan war logs so hastily has upset many of the core members. WikiLeaks' plan was to give material to newspapers only when the organisation itself was ready to publish, Daniel tells me, and they were not ready to publish the Afghan war logs at the end of July. Only at the last minute did Daniel and the others discover Julian's deal with the *Guardian* and the July publication date. A hurried effort was made to identify the most sensitive and 15,000 were held back, but no one had time to check the remaining 76,000 that were published online.

'If what you are doing is so critical,' Daniel concludes, 'then you need to be sure you are in control of what you're doing and we could no longer predict what was happening.'

For the first forty-eight hours after publication of the Afghan war logs on 25 July, the news revealed by the leaks generated a global debate about the Afghan war and America's role in it. Among the questions asked were: why was America still cooperating with Pakistan's Inter-Service Intelligence (ISI) when the reports revealed a widespread belief among the American soldiers that the ISI was collaborating with the Taliban? Was it right for Hillary Clinton to pledge $7.5 billion of American aid to such a country? What was the full extent of Task Force 373? What about the use of Reaper drones and the extent of unreported or misreported civilian deaths?

The initial reaction from the Pentagon and Washington politicians was to focus on the leak itself, claiming it endangered national security, though White House spokesman Robert Gibbs did say the billions of dollars given to Pakistan could be at risk if allegations of collusion with the Taliban were true. John Kerry, chairman of the US Senate Foreign Relations Committee, admitted that the disclosure of botched attacks by Task Force 373 and unreported civilian deaths raised questions that the Pentagon and White House needed to answer. They might have been forced to do so in the face of sustained public pressure, but on Wednesday 28 July the focus shifted.

The Times newspaper ran on its front page an article by defence correspondent Tom Coghlan:

Afghan leaks expose the identities of informants

Hundreds of Afghan lives have been put at risk by the leaking of 90,000 intelligence documents because the files identify informants working with Nato forces. In just two hours of searching the WikiLeaks archive, *The Times* found the names of dozens of Afghans credited with providing detailed intelligence to US forces.

WikiLeaks had not in fact published all 90,000 records but Tom Coghlan had gone through some of the records released after returning from Afghanistan and was disturbed to find numerous instances where informants were identified. He told me, 'One in particular was a very detailed meeting between a US officer and an insurgent considering joining a reconciliation programme. His name, father's name, place of habitation and other identifying details were included. I was very shocked by this and the other cables I had seen. I did not, and do not, consider that the decision by WikiLeaks to release those meetings was a responsible one.' The story also gave *The Times* a way to challenge the *Guardian*'s domination of this news event. His article quoted Robert Riegle, a former senior intelligence officer: 'It's possible that someone could get killed in the next few days.'

The rhetoric was ramped up further the next day in another *Times* front-page article, 'Backlash grows over "hitlist for the Taliban"', and an editorial stating:

As a result of the 90,000 secret US intelligence documents about the war in Afghanistan released on his website, innocent ordinary people will die. This is not speculative hyperbole but inevitable fact.

Apart from the continued, and incorrect, claim that 90,000 rather than 76,000 documents had been published, it was speculative hyperbole as to date no one is known to have been harmed as a result of WikiLeaks' publication of the Afghan war logs. Coghlan told me he didn't have an axe to grind, and 'the bad publicity for WikiLeaks on this issue was justified. The Afghan war logs raised important questions about the US role in Afghanistan, but I was disturbed by the attitude of WikiLeaks. Julian Assange in particular appeared to have little concern for Afghans named by his website.'

Whatever its motivation, *The Times* newspaper had aided the Pentagon and the US political establishment in shifting focus successfully from the behaviour of the US military in Afghanistan to that of WikiLeaks just as had happened with the 'Collateral Murder' video back in April. On 30 July Admiral Mike Mullen, the chairman of the Joint Chiefs of Staff, held a press conference in which he told reporters that 'Mr Assange can say whatever he likes about the greater good he thinks he and his organisation are doing but they might already have on their hands the blood of a young soldier or that of an Afghan family.'

This was speculative blood, unlike the real blood revealed to be on the hands of Task Force 373 (a unit ultimately over-seen by Admiral Mike Mullen), but the speculative blood became the story. This was the information war that Nick Davies had predicted. What he hadn't predicted was how much Julian Assange would overplay his hand to the extent that he diminished the impact of his own leaked material.

'For the first forty-eight hours we were generating a global debate and minimal aggression from anyone because our position was so unassailably correct. It was brilliant,' Nick told me. 'Then *The Times* and the *Wall Street Journal* – Murdoch titles – both started running stories saying the data was endangering lives. None of us were surprised by this but of course Julian had thought he knew best. It shifted the whole debate. By the end of that week it was almost entirely about the Washington PR line – "WikiLeaks have got blood on their hands." We were losing the information war because he'd given them ammunition to fire at us. I don't think the Pentagon could believe their luck.'

When Davies had met Julian in Brussels they discussed at length how media management was the speciality of the American government. The US had created what Davies called 'the biggest media manipulation machine in the history of mankind', with propaganda machinery in place in the military and intelligence agencies operating through a structure of public relations and press offices to feed facts and some fiction to the resource-deprived news networks.

As I see it there are two important issues surrounding the publication of controversial information like the Afghan war logs: ethics and strategy. Every journalist is at some point faced with moral and political decisions about what can be published. The ethics of publication is a vast discussion but it can be distilled down to a simple rule: what is in the public interest? This often gets confused with a politician's interest, a wealthy person's interest or the interest of the current people in power. Sometimes reporters can confuse the tittle-tattle that interests the public with journalism in the public interest. What defines the public interest is that which informs and enlightens society. The failure of journalism in Iceland, for example, was precisely its failure to represent the public interest; instead there was collusion with the interests of the powerful. The other crucial issue is strategy. How can you defend against claims that what you publish is irresponsible or dangerous?

In July, when Nick was seeing Julian frequently, he had raised the subject of the moral and strategic aspects of publishing the Afghan logs. Davies predicted that the US government's first line of attack would be that publication was irresponsible and was helping terrorists. But this line would run out of fuel after a couple of days – then they'd have to start accounting for the actions exposed by publication. The *Guardian*, like the other print publications, was only going to publish those documents it had read and checked so as not to put anyone's life at risk or give officials any excuse to claim the paper was a public menace. Julian, by contrast, planned to publish all 90,000.

It was the reporters from *Der Spiegel* who first locked onto Julian's absolutist publishing policy. They told the *Guardian* there were things in the documents that identified inform-ants who might suffer reprisals. David Leigh and the *Spiegel* reporters pressed on Julian the moral and political problems of just dumping the entire data set. Julian was very resistant to the idea of holding information back and said the people exposed were informants and as such they deserved whatever was coming to them. (Julian Assange has denied saying this but when I asked Nick Davies, David Leigh and *Spiegel* reporter John Goetz, they all claimed otherwise.)

David Leigh tried to tell Julian it wasn't as simple as that. 'If a crowd of Americans armed to the teeth comes to your house and starts asking questions you may well think it wise to comply,' he said. 'It doesn't follow that you're some kind of capitalist running dog. We thought Julian's attitude was terri-fying really. Me and the guys from *Spiegel* went pale.'

About a week before publication, Julian agreed to withhold 15,000 of the most sensitive records, including threat reports and those he'd located using word searches for terms such as 'HUMINT' and 'source'. But the hurried, haphazard approach meant many sources slipped through, and it was these that *The Times* and the Pentagon picked up on.

'It was so frustrating,' Davies said later. 'He could have been in such a powerful position if he had just listened. But Julian doesn't listen to anyone. He always thinks he is the cleverest person in the room.'

There were other problems. Nick Davies had a call the day before the *Guardian* went to press from another investigative reporter, Stephen Grey, who specialises in Afghanistan and who broke the story about the CIA's use of extraordinary rendition. 'You'll never guess who I'm here with,' Grey said. He told Nick he'd just filmed an interview with Julian Assange at the Frontline Club; not only that, he had all the Afghan war logs and a summary of all the media partners' stories. 'He [Julian] says he thinks it might rock the boat a bit round at the *Guardian* when you find out what he's done so I told him I'd call you.' Nick asked Grey to put Julian on the line.

'Stephen's just told me you've done an interview with him. Can you just bring me up to date here with what's happening?'

Julian confirmed he'd given Channel 4 the database and an interview about the stories.

'Is Channel 4 the only one?'

Julian said no, he'd given the whole database and a summary of the *Guardian*'s stories to CNN and Al Jazeera.

'When did that happen?'

'Yesterday,' Julian replied.

This was extremely worrying. The media partners had conducted the operation with the greatest possible secrecy in order to avoid the clear risk of either the US or UK government going to court in London to get an order stopping publication. All three print partners had dedicated vast resources to the project as a result of Julian's guarantee of exclusivity. Now that secrecy was blown with all the legal risk that entailed and there was no guarantee they would be the first to publish. In Norway I'd heard Julian attribute WikiLeaks' initial lack of success to its open 'wiki' format – he realised there was no incentive for a newspaper to spend time analysing huge data sets if the material was out in the open. It meant another rival publication might be doing the same analysis and be much further along. They were attracted by exclusivity.

'What you've done is very surprising,' Davies said as calmly as he could.

'This was always part of our agreement to get TV involved,' Julian replied.

Nick later reflected on how 'psychologically interesting' Assange's response was: 'I can see how he might lie about what happened to someone else who wasn't at this meeting and he might be able to persuade them. But I was there. He must have known that I would know that was a lie . . . He knew very well the need for exclusivity and that was part of our agreement. So to find out on Saturday that he'd gone and given the data not just to Channel 4 but Al Jazeera and CNN was really problematic.' Assange told Davies the TV stations had agreed to an embargo, promising not to publish until after the *Guardian*. Davies was dumbfounded by his naivety as Julian barely knew the journalists. In fact Channel 4 attempted to put out a special report that Saturday but were unable to get a slot. Only by luck, they put the story on their website at 10 p.m. on Sunday, exactly the same time as the *Guardian* and the other papers, but billing it as 'world exclusive'.

These media deals were causing problems among the WikiLeakers too. Birgitta Jónsdóttir was shocked by the cavalier way Julian had published the Afghan war logs in such a hurry. Relations between Julian and Daniel had deteriorated since their February trip to Iceland and Daniel tried in vain to find out what the plan was for the next big leak of 391,832 US military field reports from the Iraq war. Julian wouldn't tell him and instead accused him of talking to the media about the growing discontent within WikiLeaks.

That discontent escalated further on 21 August when the Swedish newspaper *Expressen* published a report that Assange was facing allegations from two women of rape and sexual assault. Julian had arrived in Sweden on 11 August as a freedom of speech hero to speak to a group of admiring Swedish activists. In an interview, he praised the country for the 'long-term support for WikiLeaks by the Swedish people and the Swedish legal system'. He stayed in the apartment of a female organiser who would go on to accuse him of sexually assaulting her. The other woman was an attendee at the conference and accused him of having sex with her while she was asleep and without a condom. How two once adoring admirers had ended up reporting Julian to the police for sexual assault and minor rape is, as yet, unclear but it wasn't the first time Julian's attitudes towards women had been called into question.

Birgitta, Daniel and the others initially supported Julian against the allegations but advised him to step out of the spotlight while the charges were investigated so as not to taint WikiLeaks' work. Not only did Julian refuse, he wanted money donated to WikiLeaks to pay his personal legal defence for the sexual assault charges. Jónsdóttir had been trying for months to arrange a time when everyone could get together either online or, preferably, in person to discuss the future of

WikiLeaks in light of everything that was happening. Julian saw this as an attack. You were either with him or against him, and with him meant following his orders obediently and without question. It was not for this that volunteers had given up their time. They wanted an alternative to authoritarian and secretive power structures, not another one with a different face.

'We had arguments before but there was always the common goal that was so strongly in focus,' Daniel told me. 'When he suspended me he said it was for "disloyalty, destabilisation and insubordination in times of crisis". These are terms from the US Espionage Act 1917. When I saw that I thought, "What the hell? Is this guy reading the wrong literature before going to bed?" Everything we'd been fighting against was manifesting itself in him.'

Daniel finally confronted Julian online:

> You are not anyone's king or god. And you're not even fulfilling your role as a leader right now. A leader communicates and cultivates trust in himself. You are doing the exact opposite. You behave like some kind of emperor or slave trader.

Julian replied:

> You are suspended for one month, effective immediately. If you wish to appeal, you will be heard on Tuesday.

The suspension was another unilateral decision. The other WikiLeakers disagreed, but Julian made it clear it was his way or the highway, telling the young Icelander Herbert Snorrason, 'I am the heart and soul of this organisation, its founder, philosopher, spokesperson, original coder, organiser, financier and all the rest. If you have a problem with me, piss off.'

Birgitta, Smári McCarthy, Herbert Snorrason, Daniel and

the Architect left, as did the majority of the Swedish and German volunteers. The only long-standing WikiLeaks supporter still onside by October 2010 was an Australian academic, Suelette Dreyfus, who had collaborated with Julian Assange on a book after his 1996 conviction for hacking. She refused to talk on the record about her role in WikiLeaks or her relationship with Julian but her involvement was confirmed to me by the aforementioned volunteers. Icelandic journalist Kristinn Hrafnsson and cameraman Ingi Ragnar Ingason also remained with Julian.

As the organisation was splintering apart in August and September, Julian was brokering media deals for the publication of the next big leak, the Iraq war logs, with another new organisation – the Bureau of Investigative Journalism in London – and several television channels including Al Jazeera and Channel 4, much to the annoyance of the original three media partners.

It was amidst this background of discontent that I came into possession of the full set of leaked US diplomatic cables. I was in regular contact with most of the WikiLeaks people, both in person and online through encrypted messaging. I'd known since the summer that there were serious problems within the organisation. The people I spoke to were unhappy about the way the video and Afghan war logs had been published, the way Julian had detracted from the strength of the material with polemic and his own personality. While they admired WikiLeaks greatly and couldn't fault Julian's audacity, they were deeply disappointed that he didn't live by his own principles. An amazing opportunity was being destroyed by the very person who had done so much to create it.

In September I met up with my source. We discussed various

things and I mentioned the upcoming Iraq war logs' publication. Julian had another big press conference planned where undoubtedly he would star. The material would also be 'dumped', so it seemed likely that attention would quickly shift from the content of the leaks to the act of leaking itself and the personality of Julian Assange – exactly the outcome my source wanted to avoid. A better solution would be to select the stories carefully and space them out over time. I began speculating about the US diplomatic cables, which I thought were the most important documents, and wondered about their impact based on what I'd read in the alleged chat transcripts with Bradley Manning.

'Do you know,' my source revealed, 'I've had those cables sitting on my computer for months.'

I'd like to think I retained an aura of cool detachment upon hearing this, but I'm fairly certain I didn't. 'No way!' I said.

'Yes, they're just sitting there.'

'Have you looked through them? Are they as game-changing as has been made out?'

'Hard to say. There's just so much there. It's almost impossible to get an overall view on it. But yes, there are definitely some interesting incidents.' (A few were mentioned and they were, indeed, very interesting.)

'What you need,' I said, 'is a good journalist to examine this data for stories. Then put forward a slow and steady stream to build relentless impact that cannot be so easily deflected or ignored. Do nothing to endanger anyone or give officials any opportunity to shift focus.'

We carried on this conversation later online. The source was worried about publishing as it might cause more problems for Bradley Manning, who was taking the blame for the leaks, but Julian had given copies of the data to several other people so there was no way of knowing what might happen. Was Julian working with someone to publish? The source didn't know but Julian had an odd habit of trusting people he'd only just met over people he knew well, so anything was possible. My source wanted the information to come out but in a responsible way; we agreed that *maybe* I could look at it on condition that I not tell anyone who'd given me access and that I sit tight on whatever I found and not publish anything until they were ready.

A little later the source agreed to let me have a look at the data on a secure website. I was faced with a vast data set and a very minimal search facility. There were 251,287 cables. It was hard to feel I'd even made a dent in the hour or so I had access. Over the next few days we did this again and I searched through the most recent London embassy cables and saw some interesting discussion about David Cameron and George Osborne. Then I looked at cables from Riyadh and the Vatican. It was tough going. Through the remote connection it was incredibly slow. There was no way to do a word search for such enlightening topics as Prince Andrew, BAE Systems, corruption or Megrahi (the convicted Lockerbie bomber released to Libya in a secret deal), because the data wasn't indexed. It would take months to sort through the cables like this. My source said there'd been a plan to build a text search facility but there hadn't been time. I offered to help. I wanted to set up a secure server, upload the data onto a private database and build a text index to analyse the data locally. The source said this needed some consideration. I waited.

At the end of September I had an interview scheduled with David Leigh. He talked me through the Afghan investigation which the reporters worked on in a secret room they had nicknamed 'the bunker', then I asked what was happening with the other data sets. This was more confidential but we'd known each other for some time and my book would be coming out long after the *Guardian* published, so he told me some of the problems they were having with the publication of the Iraq war logs. What about the cables? I asked.

'Who said we had the cables?' Leigh said, looking suspicious.

'Oh, come on. Everyone knows you have all three leaks.'

He leaned in conspiratorially. 'We don't have the cables.' It was a bold admission.

'You don't? How come? Does the Bureau [of Investigative Journalism]?'

'No, the Bureau doesn't have them either.'

'You really don't have them?'

'No.'

'How could you not have them? Even I've seen them.'

'You've seen them?'

'Only in passing.'

Leigh's eyes locked onto mine like bright blue lasers. I suddenly felt very sorry for those who'd found themselves the target of his investigative zeal.

'How did you manage that? Did Julian give them to you?'

'I can't reveal my source but it wasn't Julian. There are several people who have this data besides Julian but surely you must know this. I can't say any more.' But of course I realised I'd already said way too much. I'd simply assumed the *Guardian* had the cables.

We were on our way outside and Leigh kept hammering away at me about what exactly I'd seen.

'I've only seen a few,' I replied evasively. 'I don't *have* them, you understand. I've only seen them and my arrangement is that I'm just sitting tight on that until things are sorted out at

WikiLeaks. Once they are, then depending on what happens I'll be back in touch. But I'm not telling you anything else, so you can stop badgering me.'

But of course Leigh kept phoning and emailing. I didn't ignore him (always a bad idea if you want an investigative journalist to get off your back – that only makes us more suspicious and determined), but I told him I couldn't say any more without the agreement of my source. He'd just have to wait until they figured out what they were doing.

I continued to look at the cables occasionally but my source could only arrange the timed access sporadically at very short notice. We agreed on Friday 1 October to transfer the data to a secure machine on which I could build a text search and begin looking for stories in a more efficient way, on the condition that this arrangement was mutually confidential. This process took a couple of hours and once it was done I waded in. The scope of the data was overwhelming. It was a mass of international relations as few outside government get to see it. Many of the diplomats were excellent writers and presented the country's movers and shakers in scenes that were as gripping as a soap opera. I delved into Saudi Arabia, a country that provokes curiosity by its repressive secrecy. I found a cable from 1996, 'Saudi Royal Wealth: Where do they get all that money?' and read in fascination the details of their royal welfare system. Payments were being divvied out at that time, according to the American diplomat, that ranged from $800 a month for 'the lowliest member of the most remote branch of the family' to $200,000–$270,000 a month for the favoured princes. Grandchildren got around $27,000 a month, great-grandchildren about $13,000 and great-great-grandchildren $8,000 a month.

On Monday 4 October my source got in touch with some

improvements to make on the database. But when I logged on later to see what these were I got a shock. My source had deleted all the data. I asked online what was going on:

HB: re machine – did you delete all the data?
Source: there's a lot of pressure at the moment
HB: so you did delete them?
Source: Yes. For now. Don't get me wrong, I like working with you, but when it comes to this data I was overly uncautious. I knew it wouldn't be enough to ask you to delete it. Sorry.
HB: what has happened?
Source: I've been put under a lot of very serious pressure and I'm afraid for my security.

I'd been rather surprised that my source had agreed to give me all the data in the first place, but if there was enough trust to do so on Friday I couldn't see what had changed by Monday. I was upset and angry though I would have been a lot more of both if I hadn't made a backup copy. (First lesson I always teach my students about data journalism: always do a backup.) However, I wasn't sure what to do with this backup. I'd been working by myself the last couple of weeks, poring over the cables, looking for stories, telling no one what I was up to as agreed with my source. But now I wasn't sure if we even had an agreement or if he'd kept it. Someone must have been told about the transfer, otherwise why would things have changed so dramatically?

On Wednesday 6 October I went to see a lawyer friend of mine. His opinion was that legally I could not be prosecuted under UK or US law for possessing the cables. What I did with them was a matter of personal consideration but he advised that there were likely legitimate secrets within the cables and I would need to be careful how I published if indeed I chose to.

I didn't yet know what to do but I did want to get to the bottom of why my source had pulled the data. As David Leigh was the only person who knew I'd seen it, I figured he must somehow be involved.

We met for coffee and I told him about the source giving me the data only to delete it three days later. 'Why would he do that, do you think?' When my efforts to replicate some of Leigh's own ruthless cunning failed, I realised I had to be a bit more direct. I said I knew something was up, and asked him what he'd done.

'Forget about that. The important question is, do you have a copy?' His beady eyes were searing into mine.

'Of course,' I huffed.

'Come on!' He jumped from his seat and jogged outside, our nearly full coffees left behind. He hailed a cab to the *Guardian* offices. As we rattled up Farringdon Road to the newspaper's swishy new glass offices at Kings Place, I asked what was going on.

'This is how it is, Heather,' he said, turning to me. 'We've just been given the data by the *New York Times*. I think they got it from Daniel Ellsberg, who was given it by Julian. So you see – we already have it. You may as well put your lot in with us.'

I'd just about processed all this and was working through the questions – Why did Julian give it to Ellsberg? Why would Ellsberg give it to the *New York Times*? – when we arrived at the *Guardian* five minutes later. Again I had to hurry to keep up with Leigh as he headed up to room 4.4. As we arrived on the fourth floor he took out a tiny USB stick from his shirt pocket.

'1.6 gigabytes on this. Amazing, isn't it!'

I looked around. This was the bunker he'd told me about at our interview. It wasn't that exciting to be honest – just a small closed office with two large desks and six computers, and at the far side floor-to-ceiling windows that look down onto the *Guardian* atrium. It had up until now been a solidly male environment so I was quite pleased to infiltrate what seemed like some arcane gentlemen's secrecy clubhouse.

Leigh logged in and loaded the US diplomatic cables database, taking me through the various search fields they'd built. My inner geek was impressed. He asked me what stories I'd come across and I mentioned a cable from the London embassy about the Trident nuclear submarine, several about the controversial release of Megrahi to Libya and about Prince Andrew making an ass of himself in Kyrgyzstan.

'We're good chums, aren't we?' he offered.

'Yes, I suppose so.'

'We did a good spread when you won the expenses case, didn't we?'

I could see where he was going with this. Whatever plans he had were affected by my having a rogue copy of this database.

'When exactly did you get the data?' I asked again.

'Just a few weeks ago.'

'Why were you calling me then?'

'I wanted to know if you had them. Now look – see,' he pointed to a list of about twenty-five documents in a folder. 'We've already found all these stories. It wouldn't make any sense for you to go anywhere else and it would make all of us very upset with you.'

I was just about to go to California for two weeks to visit Google and Facebook so I asked when they were planning to publish; he said they hadn't yet decided. There were some problems with Julian but they wanted to go as soon as possible, maybe the first week of November. What about Bradley Manning, I asked – won't this make it worse for him? Leigh's response was unequivocal.

'It can't get any worse for him than it is already. What's done is done.'

I said I'd think about it. He extracted from me a promise that I wouldn't do anything until we next met.

The day before I left, on 12 October, Leigh phoned to say he'd just had a 'sinister' telephone call from Julian Assange, who was in a rage over my possession of the cables database. 'He says he knows where you and your husband live. I thought I'd better ring you.'

I wasn't that worried but I asked Leigh if he thought I should be.

'I shouldn't think he'd do anything. I told him you were in America. It's doubtful he'll follow you there. You couldn't be in a safer place.'

The Iraq war logs were published on 22 October when I was in California. These revealed several major stories. Nick Davies had discovered how the US had ordered troops to ignore abuse of detainees by the Iraqi police and they'd failed to investigate known instances of torture, rape and even murder. Davies found one log that revealed the existence of a video given to the American military in December 2009 of Iraqi army officers executing a prisoner. Despite at least one of the assailants being named, no investigation was conducted and Davies found there was a formal policy to ignore such allegations by recording them 'No investigation necessary' and passing them back to the same Iraqi units accused of committing the crimes.

The organisation Iraq Body Count identified 15,000 previously unknown civilian deaths from the logs, despite US and UK officials consistently denying that any official record of civilian casualties existed. The logs revealed 66,081 non-combatant deaths out of a total of 109,000 fatalities in Iraq. The *Guardian* reporters also discovered that the US helicopter gunship featured in the 'Collateral Murder' video had been involved in another controversial incident – killing Iraqi insurgents after they tried to surrender. A military lawyer back at base advised the pilots they could kill because, 'You cannot surrender to an aircraft.'

Again, however, focus on the stories lasted only a couple of days, after which attention shifted to Assange himself and the dangerous way WikiLeaks had previously published Afghan informants' names. To date the issues raised by the leaks have been ignored, and even the perpetrator identified as Major Ali Jadallah Husayn Al-Shamari, the intelligence officer in the video of the detainee who was murdered, has not been held to account.

Assange drew in his largest press conference crowd yet for the leaked Iraq war logs, restoring his reputation after the scandal of the rape allegations, and even seemed to be on his way to rebuilding WikiLeaks with Kristinn Hrafnsson as his new spokesman. Yet he too lost control of the media spotlight – the following day's *New York Times* featured a front-page profile of Assange which interviewed several of his former comrades, who cited what they saw as his 'erratic and imperious behavior, and a nearly delusional grandeur unmatched by an awareness that the digital secrets he reveals can have a price in flesh and blood'.

Assange was furious, 'incandescent with rage' as Leigh described it, and became obsessed with the profile. More than anything he wanted to 'punish' the *New York Times* for writing it. He would cut them out of the deal for the most important leak of all: the US diplomatic cables.

Ambush

Those who bathe in the media sunlight have to acquire a protective tan. Clearly Julian Assange had not. He had benefited considerably from a huge amount of media attention, so to have had such an enraged reaction to perceived bad coverage reveals at best a lack of professionalism, at worst a dangerously unstable ego. We might recall Danny O'Brien's earlier observation that those with power are often the last to realise they have it and continue to see themselves as a victim even while they dominate their chosen field.

The *Guardian* were not on board with Julian's vendetta. A deal was a deal and this deal had been done with the *New York Times*. It would not be right to cut them out just because Julian didn't like the profile they'd published, so editor Alan Rusbridger and deputy editor Ian Katz called a meeting on 1 November to clear the air and go through the logistics and running order of stories with *Der Spiegel*. Julian was meant to arrive at 6 p.m. but he texted Katz to say he would be late. As this was not unusual, Katz thought nothing of it. At 7 p.m. Alan Rusbridger also had a call, from Mark Stephens, a libel lawyer he'd known for years. Stephens had something to talk about and wondered if he could pop round; as Julian was running late, Rusbridger agreed.

The next thing he knew, Mark Stephens came barrelling into the editor's office like an oversized pug with Julian Assange, Kristinn Hrafnsson and junior lawyer Jennifer Robinson. They'd all managed to get through security and now here they were. Mark informed Rusbridger that he was representing Julian Assange.

'You should have seen the looks on their faces when I turned up with the lawyers,' Julian told me later. But it was unclear what advantage he thought he'd gained. There was never a written contract between the media organisations and WikiLeaks. Nothing was formal or litigable precisely because WikiLeaks operated as though it were not a legal entity. All they had was what Julian called a 'gentlemen's agreement'.

Rusbridger said he'd call in the *Spiegel* guys and David Leigh.

'No, we want a private meeting,' said Julian furiously. 'I think you've given this to the *New York Times*!'

'Look, Julian, there's a group of us here and we're all ready to talk to you. Why don't we get everyone in?'

Julian, by now practically pounding on the editor's desk, responded that no, he wanted an answer there and then.

'I know you've given it to the *New York Times*,' Julian thundered. 'Now answer me directly. Did you or did you not give it to the *New York Times*?'

'*I* haven't given the *New York Times* anything,' replied Rusbridger.

'This is a disgrace. You *did* give it to the *New York Times*!'

'You can ask this question many, many times,' said Rusbridger, 'but I'm not sure where it's going to get us so let's get the others in.'

The others did then come in but David Leigh took one look at Mark Stephens and asked what he was doing there. On discovering that Stephens was now representing Julian, Leigh was outraged. 'Fancy *you* bringing your lawyers in here! I'm not saying another word until we have *our* lawyers.' There followed several minutes while Rusbridger attempted to find either the in-house or on-call lawyer. One was on her bicycle and didn't hear the phone, the other was in the gym.

'This is either a journalistic meeting or a legal meeting,' Leigh declared. Mark Stephens and his colleague eventually agreed to leave the journalists to thrash it out.

What Julian was failing to understand, Rusbridger and Leigh explained, was that they didn't have to negotiate on his terms any more because there was another source – me.

Julian denied there was a second source, asserting the sole source was WikiLeaks.

'Either it's a second source or it's one source,' Rusbridger countered. 'If it's one source then you've lost control of it. You've breached your side of the agreement because you've given it to someone else. Which is it?'

Julian wouldn't engage with that point. He accused me of stealing the cables through 'criminal deception'. While he didn't elaborate what these nefarious methods were, he claimed to know enough about how I operated to 'destroy' me.

He turned to David Leigh. 'Did *you* give it to the *New York Times*? Was it you?'

'I'm not going to talk about it. It's all got to be done. We have to get on with it. Heather's got her own copy and she's a free agent.'

'It doesn't belong to her. She stole it!' he repeated. Then he uttered the most amazing claim of the night: he would sue both me and the *Guardian* for depriving him of his 'financial assets' as a result of my getting the leak of 'his' leak.

That's when the editor-in-chief of *Der Spiegel*, Georg Mascolo, pointed out that Julian was 'not quite right'.

'I think you are mistaken about who actually holds the title here,' he said with the slightest of smiles. 'If anyone has a legitimate claim on this information it is not you but, surely, the US government.'

* * *

The meeting at the *Guardian* on 1 November carried on into the night. The situation wasn't ideal for anyone but somehow they had to make it work. Eventually things calmed down and Julian got over his obsession with the *New York Times*, at least temporarily, and they went through the practical issues of how to produce this enormous publishing venture. Yet after dinner Julian again came back to the *New York Times*. Rusbridger said he would go and ring the editor Bill Keller. What did Julian want? A front-page apology.

Pause. 'Well, I'll go and speak to him.'

In his office Rusbridger made the call: 'Bill, I know what you're going to say to this but he wants a front-page apology.'

'Dream on. He can write a letter.'

It was midnight. Rusbridger brought back the news. Julian kicked off again. At 2 a.m. Rusbridger was done.

'Look, we all have to go home. You've essentially only got two options. You can cut us all out or we'll do it. If you cut us all out your problem is that we have it. Heather has it so you can't stay in control of this. Or you do it with us.'

He wouldn't agree and they left, exasperated and with no deal.

The next morning Mark Stephens rang Alan Rusbridger. 'We're on.' There was a short pause. 'But he wants to bring in the Romance languages.'

9

To the Brink and Beyond

Cambridge, Massachusetts, 21 to 24 November

While a group of five newspapers was preparing for the biggest leak in history and Julian Assange was overseeing the construction of a new website to publish the cables, Bradley Manning was entering his sixth month of solitary confinement. At the end of July he was flown from Kuwait to the brig in Quantico, Virginia, and in their ongoing investigation, officials from the FBI and US Army Intelligence targeted the city of Cambridge, home of the Massachusetts Institute of Technology. There were rumours in the media that co-conspirators at the university had helped the leaker with some of the technical aspects of the leaks and that Manning had visited the area the summer of 2009 and January 2010.

It appears that what piqued the interest of officials are allegations that two MIT students worked for WikiLeaks and gave Manning encryption software and guidance to transfer data from a secure server to WikiLeaks without being identified. I've traced these allegations to a single source, the hacker Adrian Lamo who turned in Bradley Manning after the alleged chat logs. I need to find these supposed co-conspirators and get them to talk to me. I don't imagine it will be easy but it isn't going to get any easier. In less than a week's time five newspapers around the world will begin publishing a series of articles based on the leaked database of US diplomatic cables, and the spotlight is sure to turn to the source – or sources – of these leaks.

MIT is America's pre-eminent technical university, known as the 'brain' of the defence industry with strong ties to the Pentagon. During the Second World War it was the single biggest wartime research and development contractor, during the Cold War more than 90 per cent of its research funding came from the Department of Defense and even in 2010 the *Boston Globe* reported that MIT was getting $750 million a year, making it one of the top five universities funded by the department.[1] Yet within its labs there is a culture quite the opposite of the authoritarian military. Here Enlightenment values dominate. Curiosity, truth and the free flow of information that leads to truth are venerated. The students have a well-established hacking culture, though 'hacking' extends beyond computers to pulling a particularly clever or bold prank like placing an improbable object, such as a fire engine, atop the Great Dome, or dressing up en masse as the character from the movie *V for Vendetta* and marching into the MIT Undergraduate Association Senate meeting to the *1812 Overture*. This might be considered radical but it hardly merits FBI attention.

I have a couple of leads: a boyfriend of Manning, Tyler Watkins, who attends Brandeis University nearby; and former Boston University student David House, the hackerspace guy we met in Chapter 2. House now works at MIT's Center for Digital Business and helped set up the website for the Bradley

1. http://tinyurl.com/6jlkj5c

Manning Support Network. Earlier in November, he was stopped at Chicago's O'Hare Airport by Homeland Security officials who seized his electronic equipment and questioned him about his visits to Manning in the brig and about WikiLeaks.

I can count on one finger my Boston contacts. Fortunately that person is Aaron Swartz, who's in the Cambridge tech/activist scene. He describes himself as a writer, activist and hacker and at twenty-five his CV is impressive: currently founder and director of a democracy campaign group, Demand Progress, he previously co-founded Reddit.com (a website for sharing news links) and was part of the original team to launch Creative Commons. At fourteen he co-authored the Really Simple Syndication (RSS 1.0) specification for publishing news updates. In the information war he's participated in a few guerrilla campaigns which have accorded him his own FBI file (posted on his blog). In 2008, he hacked into a federal court library system to leak over 18 million public documents that the government had been charging citizens to access. Swartz only realised how much trouble he was in when the FBI started monitoring him. He got himself a lawyer, but luckily the *New York Times* got on the case and made him something of a cause célèbre. The FBI eventually backed off: it looked bad to spend taxpayers' money going after a kid for making public records more publicly available.

Aaron has set me up with a room in a place called the Acetarium but even standing outside the door on this cold November night I can't tell if it's a hostel, a hotel or a house. I telephone the proprietor Benjamin Mako Hill and in a few minutes I see pale legs jumping down the stairs. He's known as 'Mako', he tells me, and he has an impish, Irish look with a pointy Pan-like beard and big mischievous blue eyes with a ring through his left eyebrow. He's wearing an American flag do-rag and a yellow cycling jacket. He's brimming with energy and hops up the stairs two at a time. On the landing is a sign: 'Shoes and pants off please'. I leave mine (shoes that is) at the door and head in.

Inside, over some home-made vegetable dumplings, I meet Mako's wife and some of the other residents: a twenty-year-old couchsurfer from North Carolina, a freelance software programmer in the spare room and a guinea pig whose owner has gone travelling. Mako himself is a scholar at MIT's media lab specialising in sociology and online communities and he's an active member of the Free Software Foundation. He sounds exactly the sort of person who can put me in touch with the people I need to talk to, but when I start asking questions he clams up. 'I'm not into that scene,' he says tersely, tapping his foot. 'I don't know any of those people.'

Later that evening, Aaron comes over to the Acetarium and tells me this used to be the original Reddit offices. He passed them to Mako when Reddit was bought by Condé Nast and he and the other founders moved out to San Francisco to live the dream. He says California wasn't all it's cracked up to be. Neither was the office job at Condé Nast. He's since been fired, dropped out of Stanford and is now a fellow at the Center for Ethics at Harvard University as well as running his campaign group. He has an intense curiosity that lasers into whatever happens to interest him at any given moment, but the attention is short, and soon he's off delving into something else. Fortunately his immediate interest is my 'quest', so he grabs a nearby laptop to see what he can find online. A quick glance of Tyler Watkins' and David House's social networks reveals they're both linked to someone called Danny Clark. It's a long shot, but I ask Mako if he knows Danny Clark. His response is straightforward enough: 'Never heard of him.'

'But he's on your list of LinkedIn contacts,' says Aaron, now perusing Mako's profile, and I remind Mako there's no privacy on the Internet. He reiterates that he's 'not involved in any of this, and I don't want anything to do with it'.

'What's wrong with answering her questions?' Aaron counters.

'You don't understand, there's been all kinds of people round here.'

'I understand completely. I was investigated by the FBI, don't forget. That doesn't mean you can't talk. We're not in a police state yet.'

I decide not to press my host any further, but I'm struck by his guardedness. Clearly people are scared, and I begin to worry if I'll get anything at all out of this trip. Maybe to make up for his reticence, Mako invites me to come along to a pub in Harvard Square where every Sunday he organises a social evening for a group of techie friends studying or working at MIT or Harvard. I meet all sorts of interesting people including a woman working on the human genome project, but the most interesting of all is another Brit who tells me he lives with Danny Clark.

It's one of those moments when I just can't believe my luck. I ask if I can come round tomorrow; he says he doesn't know if Danny will be there, but he can't see why not. Just to be sure I get his address and phone number before the end of the evening.

Later, as Mako and I head home, we pass the Harvard Educational Institute where Mako studied quantitative sociology. He's optimistic about the Web, as it opens up new platforms for publishing, such as a project he recently completed at the MIT Media Lab called 'Beyond Bars' that gets prisoners blogging.

Does he consider himself a hacker? Definitely not, he says. 'I'm more about building products to enable and empower people, not finding ways to shut them out.' He thinks the hacker community is comprised of a bunch of alpha geeks – macho, misogynistic, thuggish – not so different from the military, really. 'A lot of them are just selfish teenage assholes. Most grow out if it, others go on to do computer security.' Even within the tech community there are different views about the definition of 'hacker'. Mako wants to focus on constructive projects so he's directed his technical skills and campaigning zeal into the free software movement. The question of who controls technology is, for Mako, 'one of the most important for the next hundred years. It's a political question about who controls our experience. That's why I've devoted so much of my life to free software.'

The next morning I'm on Chestnut Street. Pika House is a three-storey wooden Victorian building, painted dark turquoise, and looks like a frat house. Luckily, Danny Clark is around today. I meet him round the back of the house where he's transforming an old bicycle shed into an office. He's in his twenties but has thinning hair, an awkward stance and glasses. We go inside the main house where there's an enormous kitchen with industrial ovens, mixers and a rack of pots and pans; on the door is a list of who among his twenty-five housemates is on cooking duty each day. A melted plastic plate hangs from the ceiling to commemorate one (British) student's incomprehension at how a toaster oven works.

I tell Danny I'm here to find out about Bradley Manning and why federal officials are so interested in MIT students. I'm not sure how he'll react but he starts right off, speaking in quite a loud voice about how he was questioned and followed in such an intrusive way that he had to find a lawyer. Yes, he says, he is friends with Bradley Manning and he came to visit at Pika House a few times.

I tell him there's a rumour going around that if Bradley Manning was the source of the leak he was helped or groomed by MIT people. What does he think of that?

People often react to probing questions defensively, even aggressively. I worry that Danny might shout or yell or throw me out.

Thankfully he does none of these things. Instead he lets out a roaring laugh and claps his hands.

'That's as good as some of those 9/11 conspiracy theories! What, are we like meant to all be secret agents of WikiLeaks here? How would the head of WikiLeaks find some random person in army intelligence who happens to be sympathetic? It's insanely ludicrous. It's become a running joke here in the house.' As for MIT being a hotbed of political radicalism, that's completely off the mark too. 'MIT did have a free culture group about copyright at one point. That was a bit more political but it totally petered out. The groups that survive here are ones that are highly technical. Not political. I have a friend who tried to organise a rally on Bradley Manning Days of Action and the number of people was negligible.'

Why is MIT being targeted? I ask. He says the government doesn't have any evidence and so they're grasping at straws, trying to find something, anything. On one occasion, for example, he posted a moan about theft and water damage at a local storage company to an online email list. Manning was on the same list. 'The Feds found out about this. They spent an entire day and a half going to local storage facilities looking for a unit in Bradley Manning's name. What were they expecting to find? It just baffles the mind.'

But there's clearly a connection, I say – how does Danny

know Bradley? He explains that he met Tyler Watkins when they were organising an event in Rhode Island. 'We drove down together,' he says. 'I left a book of mine in his car, *Surely You're Joking, Mr Feynman*, and when Brad was next visiting Tyler he saw the book and asked if he could borrow it. Tyler told him it belonged to me and Brad got in touch. I think he was just so excited to find someone interested in the same things, I guess.'

Danny visited Brad at Pika House in September 2009, taking numerous photos which he posted on Facebook. Maybe the FBI were getting their leads from social networks, too, as they were keen to question not only David House but everyone who lived at Pika. 'The person who was following us seemed pretty clueless,' he recalls. 'The government can't do domestic surveillance so they hire some guy to do it and they don't even know he's inept because they're not supposed to have hired him.' In addition to 'stocky balding man', as they called their most frequent follower, Danny says there has been an 'alphabet soup' of officials asking questions and doing surveillance in the area, including the FBI, the army intelligence Criminal Investigation Division, and Defense Intelligence. In the end Danny's lawyer had to send a letter to multiple different agencies ordering them to stop harassing his client.

Danny believes the allegations about co-conspirators come from Adrian Lamo, who has astroturfed the idea in response to a backlash from hackers who think him a snitch for turning in Bradley Manning.

Danny is one of a handful of visitors allowed to see Manning, who he says is doing surprisingly well.[2] He's in high spirits,

2. This would change. By late December both David House and Manning's lawyer reported that Manning's physical and psychological health were failing after months in near-solitary confinement.

'probably because they have him on various SSRI [anti-depressant] drugs'. He says they don't discuss the case as all conversations are recorded. 'Mostly what I do is track down people for him and get them on the [visitor] list.'

Manning's other regular visitor is David House. 'As far as I know,' Danny muses, 'his involvement is even more tangential than mine,' as David only knows Brad through him. I ask if he wouldn't mind phoning David House and Tyler Watkins to see if they'll talk to me. Watkins doesn't answer and Danny isn't hopeful he'll talk. 'He's not talking to anyone these days, even me. It's awkward. He has a lawyer, I have a lawyer . . .' But David House has said he'll think about it.

A couple of hours later while I'm looking around campus, I get a call from David. 'Come round to the Pika House. I'll meet you there.' When I first see him he's hammering out a copper pipe from the breeze-block of Danny's reconstructed garage/office. We end up having a five-hour conversation that begins along the streets near Pika House, through the underground tunnels of MIT, and finishes at a restaurant. He's intense, full of energy, very smart and focused. I tell him I'm here to get to the bottom of this story about Manning having co-conspirators in the MIT area.

'The reason you've heard that story and the reason that story has been in the media – the *Boston Globe*, *New York Times*, CNN – is because of Adrian Lamo. Adrian Lamo has never been a hacker – he's a media manipulator. Adrian plays to the romantic noble hacker narrative and he's made a lot of media contacts but what he's doing is part personal vengeance and part personal narrative. As far as I know Manning had no handler in Boston. The actual truth isn't as appealing as this story of Lamo's. Manning was here because he was lonely. He was a charming but alienated individual.'

House says he's on the visitor list after initially going to support Danny then found he liked Brad. 'We share similar ideas about information warfare, transparency, all that. We've read a lot of the same books.' If Manning is the leaker he's not saying. 'We don't discuss the case because he's under constant surveillance.'

House says he's known Lamo for five years, long before the Manning situation. Lamo came to Boston in March 2010, wanting to be hooked up with the hacker scene there, but had little success. 'I told him I couldn't take him anywhere, because he was wandering into poles and things, falling over, they almost wouldn't let him in my dormitory because they thought he was drunk. I told them he was mentally ill. It was embarrassing and I told him very politely that I couldn't introduce him to anyone. Lamo and I haven't talked since.' Lamo has admitted publicly to being addicted to various prescription and non-prescription drugs and being diagnosed with Asperger's disorder.

'When the *Wired* story broke about Manning,' House continues, 'Lamo was very unhappy with me. He thinks I've turned the Boston hacker community against him and he doesn't like this area because of what happened. The narrative that Manning had a handler – it means WikiLeaks is an espionage organisation, that it actively solicits leaks. If Lamo can sell this narrative to the army and the media then he's a national hero, he's exposed a spy. Otherwise he's a snitch and ignoble and a traitor. Whatever narrative he can sell that relates to espionage makes him look better.'

In response to House's comments, Lamo said he knew both Manning and House from the Internet: 'I've only met House once, and consider myself a friend of the Boston hacker scene, having been born there. I don't know of any disagreement with him.' Lamo added that evidence is with the Department of Justice and Criminal Investigation Division and was sufficient to produce subpoenas for a grand jury investigation.

Having the FBI and army intelligence Criminal Investigation Division hanging around campus must have had an effect. People seem afraid. House says it has served only to radicalise the students. 'They feel the government is declaring war on them.' It's certainly the case from the people I talk to during these three days that there aren't enough pro bono lawyers for all the tech people being questioned and harassed. None has been charged. House believes this will end in one of two ways: either the US reforms and returns to its Enlightenment roots or all the activists, 'the cream of humanity', as House calls them, are rounded up by a surveillance state.

It's a dire prediction but these are the outcomes as seen in the context of an information war. Either we believe in Enlightenment values and let nothing stand in the way of truth, or we have conviction bred of faith and fear. In scientific (or any other) investigation the biggest danger is self-deception. It's bad enough deceiving others (making the facts fit a preconceived theory or desire) but worse to deceive oneself. I'd come to Boston with a preconceived idea. The facts were proving a little less sexy than the co-conspirator story I'd imagined. Maybe the people I spoke to were holding something back. Maybe there were others I'd missed, but the evidence I'd seen indicated that the rumours were exactly that. On the plane back to London I thought about all the stories we were going to publish as a result of one enormous leak. I thought about the possible motivation for a person to leak so much classified information. There was no allegation that Manning was a spy or selling secrets to the Chinese or Russians. In his 'confession' he says, 'I want people to see the truth . . . regardless of who they are . . . because without information, you cannot make informed decisions as a public.'

Then I recalled something that Richard Feynman (that famous physicist and MIT graduate whose book Danny and Brad had been reading) wrote about the NASA space shuttle *Challenger* disaster which he'd been tasked to investigate as part of the Rogers Commission in 1986:

> NASA owes it to the citizens from whom it asks support to be frank, honest, and informative, so that these citizens can make the wisest decisions for the use of their limited resources. For a successful technology, reality must take precedence over public relations, for nature cannot be fooled.

I wonder if Bradley Manning had read that and substituted for 'NASA' the words 'the US government' and for 'technology' the word 'war'?

When I arrive back in the *Guardian* newsroom on Friday 26 November Ian Katz the deputy editor has gone from chipper to haggard. The Romance languages that Julian wanted on board have meant the inclusion of two more newspapers: *Le Monde* (afternoon French daily newspaper) and *El País* (Spanish daily). Needless to say, this has created an extraordinary logistical nightmare for all those involved. There's continued anxiety about Julian Assange. After the ambush meeting, David Leigh wants nothing more to do with him so Ian Katz has become the point man, rushing around trying to keep Julian happy so that you might think Mariah Carey was fronting WikiLeaks. 'Now he wants to give all the data to Iceland and other places on Sunday,' Katz says, shaking his head.

The reason they all dance to Julian's latest demands is because they're petrified he'll dump the entire cable stash. To do so would be brazenly detrimental to WikiLeaks' own reputation, but in Leigh's words Julian is, 'to use the technical term, a dangerous lunatic'. And Julian has a track record of dumping data (with the Afghan war logs) and doing side deals, so they're continually kept on their toes.

Yet despite all this, relations are better than before. After the meeting on 1 November, Julian agreed to abide by the original agreement and on the 11th the editors and project leaders from the five papers came to Alan Rusbridger's office to hammer out the details of publication. I was there too as I'd formally agreed to join the *Guardian* team a week earlier – it made sense as I knew them and they already had the data from the *New York Times*. But I was still confused about how the *New York Times* had got it. I asked Ian Fisher, the deputy foreign editor of the paper, during a break, who gave them the data. Fisher said he'd promised his source confidentiality and couldn't say.

Both David Leigh and I opted out of the 11 November meeting with Julian Assange. David claimed to have the flu, and Alan Rusbridger thought seeing me would set Julian off – 'It's bad enough having the *New York Times* here.' While the prepublication drama seemed to have dissipated for now, so too had my faith in Julian Assange. After a brief online encounter where I confronted him about his threats to sue me for 'criminal deception', we met after the meeting. He was thinner now, coughed continually, and his platinum bob was replaced by a hatch of black and blond spots. I'd hoped we could clear up what had happened but our exchange didn't go well. He made a number of personal attacks and accusations about my trying to sabotage his enterprise; it seemed to me he was just furious about losing control of the cables. I'm sure the Pentagon felt exactly the same way about him. Plus he was no longer the centre of the story – it was the newspapers and journalists that were doing all the work.

Later, I contacted my source to ask about these new allegations made by Julian. They explained that three days after the upload of the data, one of the WikiLeakers still loyal to Julian, Ingi Ragnar Ingason, had come to visit. They'd gone outside where my source was told a story that I'd been selling the data around Fleet Street and had divulged the source's identity (both utterly untrue). This was obviously very upsetting and the source felt betrayed. They were given no opportunity to contact me to check the allegations and this bluff succeeded in getting the source to confess that the transfer had occurred. The source was told if they didn't cooperate immediately there would be trouble; threats included prosecution and prison. As a result, the source agreed to pull back the data and sign a letter stating that the information was taken through deception.

'But why didn't you come to me with this?' I asked in dismay.

'From my perspective it's pretty simple – they're in a position where they can really do me a lot of damage. In fact they already have. I don't know how real that threat is, but it's certainly bearing down on me like a ton of bricks right now.'

I thought of Mako and the others at MIT. Fear and threats had won the day yet again.

Back in the *Guardian* bunker, on the Friday before publication, Katz told me they had had a visit from 'the Americans'. Despite all the doomsaying from their lawyers, the *Guardian* had not faced any interference from the American government as a result of the Afghan or Iraq publications but the State Department was extremely worried about the diplomatic cables. The chargé d'affaires and head of public affairs visited Rusbridger. Then they had a conference call with P. J. Crowley, the State Department spokesman at the time, people from

Hillary Clinton's office and members of the intelligence community. 'They wanted to know which cables we were going to use,' Rusbridger told me. The *New York Times* had given them a full list but the *Guardian* was using many more. While Rusbridger refused to reveal which cables they were going to use, he did give the American officials an indication of the areas they would be covering. 'Why don't you tell us what you're worried about?' Rusbridger asked. They listed Yemen and American special agents in Pakistan, then asked again for the list. Again Rusbridger said no, that wasn't how European newspapers operated, but he invited Crowley to get in touch if they had concerns with the coverage. Crowley gave Rusbridger his email address. 'Will you be there over the weekend?' Rusbridger asked.

'Oh, yes,' replied a vexed and frustrated Crowley. 'We'll be here all weekend.'

National security

Had anyone asked Crowley the following Monday why he'd had to cancel his weekend plans, the answer would have been unequivocal: national security. It's the reason most often cited by governments to keep secrets from their citizenry. It exists to protect, but who needs protection and who is being protected by official secrecy?

The US government had adroitly turned previous leaks such as the 'Collateral Murder' video and Afghan war logs into debates about WikiLeaks by claiming that the publications were a danger to national security; they claimed that the video was kept secret for the very same reason. Yet the leaked footage did not endanger anyone's life or harm existing operations, the twin exemptions to official commitments to transparency.

As it stands, national security is only marginally about the protection of the general public. Primarily it is in place to protect a 'nation', but in practice this means protecting the status quo, since any attempt to change the status quo can be seen as a threat. The problem here is that such a definition allows for little change, as any protest or form of civil action can easily be clamped down upon under the guise of protecting national security even when such protests are precisely what the public want and need.

Libya is a nation, as are Egypt and Tunisia. Protestors who demanded an end to these countries' autocratic regimes in early 2011 were cited as dangers to national security, for good reason, as they were explicitly trying to change the structure of those nations. But is this a crime? Isn't it better for the long-term benefit of the people for these despots to be overthrown? The answer is clearly yes if principles like rationality and human rights are the goal, but not if security is paramount. Then stasis is preferable, even if it means that a large number of people must live under a corrupt dictator. Freedom requires a tolerance for risk. As soon as we lose that tolerance we are on the path to losing freedom. It was for this reason that Benjamin Franklin stated that 'those who would give up essential liberty to purchase a little temporary safety, deserve neither liberty nor safety'.

In his book *The J Curve* Ian Bremmer uses American sociologist James Chowning Davies' J Curve theory of political revolutions to describe why nations rise and fall. He argues that to progress from a closed regime to an open one, a country 'must go through a transitional period of dangerous instability'. There is no guarantee of success as there are many who will use the instability as an opportunity to 'game the system' and fill the power vacuum before civic society is strong enough to provide a check on power. A fully closed country can be stable in the short term though not nearly so stable or prosperous as the fully open society. Bremmer points out that if stability is the overriding goal then it is far simpler and easier to have a closed, authoritarian system. A closed system remains stable because it isolates its citizens from each other and/or the outside world, or it is stabilised in some way by the outside world (as Saudi Arabia is supported through its oil wealth by America). Yet the benefits from an open society exceed those of any closed society in the long run. To make it to the heights of the J curve requires a nation to hold firm to the ideals that made it open in the first place. Too much focus on security will send the country back down the scale. This is most likely to happen in response to real, perceived or imaginary threats. Fear is the essential element in an authoritarian society.

There is also the question of why the organisational structure 'nation' is deemed more legitimate than any other grouping. National security doesn't allow for such questions to be asked, or for any reconfiguration in the structure of 'nation'. Israel is a nation created where there wasn't one before. The USSR is an entirely artificial construct kept together in places through force. The concept 'nation state' is a power structure but it has little to do with popular consent. Many nations are monarchies or indeed dictatorships – why should a dictator have more legitimacy in the world than a collection of people who have banded together by consent? Even a global organisation such as the United Nations is stymied by its adherence to the 'nation' as the only legitimate grouping of humans. One only has to look to the 2010 election of Iran – a theocratic state that enshrines stoning in law and lashings for women judged as 'immodest' – to the UN's Commission on the Status of Women to see that nations are not the best representatives of the public's interest. The Unrepresented Nations and Peoples Organisation (UNPO) was set up in 1991 precisely to address the indigenous peoples, minorities, and unrecognised or occupied territories who are not represented as a 'nation' yet nonetheless want to protect and promote their rights and their environment, and engage in diplomacy. Their many members include Chinese separatist movements such as Tibetans, and Kurds in Iran and Iraq.

One reason the US has put forward for not joining the International Criminal Court is that the body places on equal terms democratic nations with those run by despots and dictators. It is easy to see a situation where dictatorships could gang up on democratic countries in such an organisation. The United Nations similarly gives the same standing to China, Saudi Arabia and Bahrain as it does democratic countries. Then there are variances between the second group – the Russian Federation, for example, is fundamentally structured as a multi-party representative democracy, but in practice is a one-party autocracy. This prompts the question: how good a democracy is needed to obtain political legitimacy?

It's easier to see when other countries abuse the term 'national security' to keep their people ignorant. The US State Department criticises China, Iran and Libya for using 'national security' as a means to criminalise, arrest, detain and sometimes kill pro-democracy campaigners. Yet when these movements happen domestically, they are more challenging. All governments have a vested interest in protecting the status quo and that's as true in the UK and USA as it is for non-democratic countries. We don't as a rule kill our protestors but they are often criminalised and certainly kept ignorant through official secrecy.

It comes down to how a country defines 'strength': to some it is the ability to crush dissent, to others it is the ability to tolerate it. America's founders wrote the Constitution to forge a country based on the latter. Yet in 2010 the Protecting

Cyberspace as a National Asset Act was going through Congress, which would have given the President power to switch off the Internet entirely – exactly as many dictators would do in 2011. This bill failed but was reintroduced in 2011 as the Cybersecurity and Internet Freedom Act minus the kill switch provision but with all the other authoritarian powers – such as filtering, blacklists and blocking – included. In addition, the Combating Online Infringement and Counterfeits Act (COICA) was introduced in September 2010 and although it had not passed into law it was used by the US government to seize domain names both in the US and overseas.

Whether new laws are even needed is debatable as the government was in 2011 shutting down worldwide websites through customs enforcement laws. The most noticeable example was the seizure of several .coms of overseas poker websites, video streaming and music blogs. All this is done without due process – only a warrant signed by a magistrate clerk, not even a judge – and the site owners are given neither notice nor a chance to mount a defence. It's an example of one nation's attempt to supersede its national boundaries and it could be the beginning of the end for Internet freedom.

This shows the inherent problems of using a legal system based on national geography in a global, digital world. The American solution has been to expand its influence and jurisdiction globally, and yet there is, as yet, no worldwide judicial accountability or oversight. In a similar way the Pentagon recently established a new command centre, USCybercom, with almost no public debate or oversight. This is part of a worrying trend in which due process and the rule of law are abandoned in favour of the authoritarian argument that the ends justify the means.

The US is by no means the only country seeking to regulate

what is said on the Internet. Across the world, governments want authority over our communication with each other and their justification is that they are protecting us. In April 2011, Russia's main security service warned that uncontrolled use of Skype and Gmail was a 'security threat', and a state tender went out for research into 'foreign experience in regulating' the Internet. Increasingly, the security services in many states are looking to China as a model for how to clamp down on Internet freedoms. These restrictions are often brought in under the mantle of protecting women and children: to stop child and other types of degrading pornography. Yet child abuse existed long before the Internet. Yes, there are bad things on the Internet but it is a communications technology. It does not have morals. It is the actions of humans to which we can ascribe morality, and if we want to punish criminal activity then let us do that, not punish the world for daring to communicate.

We should also note that abuse is most rampant not in open societies valuing free communication but in the opposite: closed, secretive, authoritarian societies such as the Catholic Church. Abuse happens not because of open communication channels but because of the way power is organised. In a top-down hier-archy where those at the top have more rights and privileges than those at the bottom, abuse will, and clearly does, occur. Getting rid of free speech and a free Internet may stop us knowing about abuse but it will certainly not get rid of it. Instead it gets rid of the evidence of how destructive authoritarian society is for many people, particularly women and children.

Publication night, Sunday 28 November

The lead-up to publication was spent in the small, white room on the fourth floor of the *Guardian*'s offices. As the weeks progressed, various foreign correspondents came to study cables from the countries where they were based. A few stories were coming to the fore: the pan-Arab concern with Iran's nuclear capabilities, the Americans' surveillance of the UN, and corruption. Lots and lots of corruption. Ian Katz would come bounding in wearing a daily changing rainbow selection of pullovers. He was now the lynchpin of the multinational journalistic operation, communicating with the five papers across three time zones through Skype and phone, and with Julian and the WikiLeaks group over the instant messenger service Jabber.

Publication was set for 21.30 GMT Sunday 28th but ironically the slippery nature of digital information was to foil even the world's most digital-savvy newspapers. A rogue copy of *Der Spiegel* went on sale by mistake at a railway station in Basel, Switzerland, at 11.30 a.m. that day. A local radio reporter picked up a copy and began broadcasting some of the stories. Then an anonymous tweeter, Freelancer_09, started tweeting them. The editor of *Der Spiegel* managed to convince the radio station to hold back but the tweeter would not be suppressed. He or she went from having 40 followers to over 600, many of them other journalists who asked him or her to start scanning in the articles. So it was that even the editors of the mainstream media found themselves powerless before a lone individual hooked to the Internet. The carefully constructed embargo was trashed and the papers rushed forward their online publication. The *Guardian*'s splash went online at 6.13 p.m.

It was later that evening, as we were all still buzzing with excitement, that another leak occurred. I discovered from the *Guardian*'s comment editor that the paper had been in possession of the cables database since August, not since October as Leigh had told me. In its coverage, the *New York Times* stated only that the cables 'were made available to *The Times* by a source who insisted on anonymity' and that they were given to the paper by an 'intermediary'. Down at the bar at the end of the night, I asked Leigh and Rusbridger about this.

'So you had the cables all along. You said you didn't.'

'Did I say that?' says Leigh, mock innocence.

Rusbridger immediately jumps in. 'Surely you've had to stretch the truth to get a story?'

I don't believe I have, but then I'm not under the relentless competitive pressure of filing daily stories (and when I was, that was in America where information was more readily available). I'm annoyed but I also understand why they did it – to lock down the story. But why did they have to lie to me?

'Well . . . we didn't know what you'd do,' says Leigh.

I wonder how far it is right to go along this route, with the ends justifying the means – surely there comes a point when you are no different from the people you are exposing? I put this to David Leigh and he says the difference is that he knows what's true and what's a lie. He might have to be a little economical with the truth to get what he needs, but once he has it he's happy to tell me the full story.

He got the cables in early August from Julian. The initial agreement had been for all four packages, but Julian had held back the cables and Guantánamo files. Then after the Afghan war logs were published Julian decided suddenly he wanted to postpone the publication of the Iraq war logs by several months.

'We were pretty annoyed with Julian by that time,' Leigh remembers. 'We'd done a lot of stuff for him, so I asked what he was going to do for us. I was back and forth to that little mews house he was staying in behind the Frontline Club. He'd been going on about these cables and how they were going to change the world and all this. I needed to see them and we needed time to go through them to see if that was true.'

Eventually Leigh persuaded Julian to give him the data. But he had to promise not to share or publish anything from the cables or even reveal he had them without Julian's permission. In September, I did my interview with Leigh.

'When you mentioned this at lunch I woke up. Jesus! I was startled out of my wits. After the enormous fuss Julian had made about how secret this all was – and the idea that you got hold of it was appalling.'

I remind Leigh that I didn't actually have the cables at that point. But Leigh figured it was 'only a matter of time' before I did. 'I brooded about this rather a lot. It changed everything. If you really had got your hands on it then it was going to be uncontrollable.'

The papers didn't like the way Julian had decided unilaterally to cut out the *New York Times* over the profile. Punishing journalists for unflattering coverage by withholding information is a common tactic of politicians. The pride and reputation of journalistic independence comes from being able to withstand this pressure and manipulation. So it put the *Guardian* in an awkward position. To abide by Julian's diktat would mean they were no longer independent. They were also of the opinion that a deal is a deal: the *New York Times* were in at the beginning so they'd be in at the end. David Leigh decided that because Julian had lost control of the information he could no longer dictate terms, and so Leigh gave a copy of the cables to the *New York Times*. The day after the cables were published, on the 29th, *New York Times* editor Bill Keller confirmed in public that the *Guardian* had been their source.

There was something of a rapprochement then between Julian and me. His anger was concentrated on David Leigh and Alan Rusbridger, whom he called 'absolute snakes'. If he thought they were bad, I said, he hadn't seen much. They were after the story and one had to operate with that understanding. If he hadn't been so antagonistic then we could have sorted this out. But nothing was ever Julian's fault and judgement fell heavily on the *Guardian* – they'd gone from being the best paper to the worst, much as Sweden moved from being a country with the world's best legal system to the worst, all based on how it had served him personally.

Once the publication was up and running the challenge internally was to maintain trust between everyone. Miscommunication was the biggest problem. WikiLeaks published a cable that precipitated the *Guardian*'s Pakistan revelations; the *New York Times* threw everything into one story, which meant the *Guardian* had to pull forward some articles they'd scheduled for later in the week. There was also the constant, looming worry that somehow an unredacted cable would get out and place someone in danger, making the papers vulnerable to government claims that it was dangerous for any outsiders to have access to all this knowledge.

Suspicion of Julian was constant due to his history both of bringing other people in at the last minute and of his general unpredictability, but these were mostly sorted out by talking directly to Julian and a few others in the WikiLeaks team. 'There was remarkably little cheating that went on,' Katz said afterwards. Most nights ended with a few bottles of Chablis. It was exhilarating and the adrenalin was such that all personal quarrels were pushed aside as everyone united to make this unprecedented journalistic project work. We felt we were part of something that was bigger than each of us. For a journalist, these are the stories you live for, and for a campaigning journalist there was the added fascination that suddenly ideas about freedom of information and secrecy were going mainstream around the world.

'There was a period of complete fury in the middle,' Rusbridger recalls. 'I can't think of a time when there was ever a story generated by a news organisation where the White House, the Kremlin, Chávez, India, China, everyone in the world was talking about these things. All over the world there were board meetings and intelligence meetings about this. I've never known a story that created such mayhem that wasn't an event like a war or a terrorist attack. There were so many issues that exploded out of this.'

While the newspapers faced relatively muted criticism, WikiLeaks came under tremendous pressure even though this time they were only publishing the carefully redacted cables supplied by the newspapers. Echoing the tactics used by Julius Baer Bank, those wanting to stop the site targeted third parties. When distributed denial of service (DDOS) attacks shut down the servers, WikiLeaks moved to cloud servers hosted by Amazon. Some visualisations of the data were hosted by a Seattle company, Tableau. Both came under political pressure to drop WikiLeaks: US Senator Joe Lieberman intervened with Amazon, and Tableau also cited indirect pressure from the US government. Both companies dumped WikiLeaks. The American-owned name registry company EveryDNS followed suit, meaning users could no longer find the website by typing in the text address wikileaks.org. PayPal, Mastercard and Visa suspended all payments to WikiLeaks and later Bank of America announced it would not process transactions 'of any type that we have reason to believe are intended for WikiLeaks'. Even Apple pulled a WikiLeaks App from its store. It was not direct government censorship but it was sobering to see where and exactly how the US government was able to control speech on the Internet.

In their fear, government officials dropped the fancy rhetoric about a free Internet and instead their power to control it was exposed for all to see. Not all liked what they saw. A backlash occurred when hundreds of mirror sites appeared for WikiLeaks. Saturday 4 December was chaos – the site would be shut down on one location only to appear somewhere else. The Internet activist group Anonymous launched 'Operation Payback', targeting the websites of those companies that had tried to shut down WikiLeaks. There were some bizarre ironies. Pakistan's High Court refused to block the WikiLeaks website while the US Library of Congress did. Meanwhile Julian's fame was reaching its zenith. He did a live Q&A on the *Guardian* website and it crashed under the strain.[3] He was on the cover of *Time* magazine in an iconic photograph that was taken up by people around the world like that of a new Che Guevara.

The US Justice Department had announced that it was opening a criminal probe into WikiLeaks and was considering laying charges against Julian Assange under the Espionage Act of 1917, though such a charge would be difficult to prosecute due to America's First Amendment protection on free speech. Plus espionage required proof that Julian had conspired with Bradley Manning or others to commit the leaks. Then criminal charges of a different nature appeared. Interpol issued a red notice on 1 December for Julian to be questioned in Sweden in relation to the sexual assault allegations from that summer. A warrant for his arrest was issued and on the 7th Julian turned himself in at the local police station. He ended up doing a week's jail time at Belmarsh prison before bail was granted and he went to stay in the manor house of a wealthy supporter.

3. As of 13 April the *Guardian* had published 287 articles on Julian Assange and 866 on WikiLeaks.

On 23 December the *Guardian* drew a line under their exclusive publication of the cables. They would still write articles but WikiLeaks was now free to share the data with other newspapers, which they did, bringing in newspapers in South America, India and Israel among others.

No one knew what the consequences of the biggest leak in history would be. The same officials who had warned that publication of the cables would jeopardise national security now tried to diminish their importance by saying there was nothing new. But it was certainly new to the people of Sudan, for example, to find out their president, Omar al-Bashir, had siphoned as much as $9 billion from that poor country to London banks, as one of the cables recounted. If the threat of Iran was serious enough to unite the Americans and Israelis with the Arabs, surely the people of the world should know too. Weren't American citizens allowed to decide whether or not their government should do business with corrupt dictators or murderous tyrants? How could they make that informed decision without information about the countries their government was doing business with?

As I left the *Guardian* newsroom on a snowy December night I paused to check the headline board at reception. There were violent protests in Tunisia. I remembered the cables about the opulent corruption of the ruling family there: while the people suffered in a police state, one of the sons-in-law of the President flew in ice cream from Saint-Tropez and kept a pet tiger in his palatial compound. Now the truth was out there. What would the people of Tunisia do?

Is this the watershed at which all secrecy becomes impossible, or the start of a new counter-attack on free speech? Certainly there are many companies and governments who, in the name of security, want to take us back to the Dark Ages.

Lock down information. Maintain control. Leaks have happened before. They are not new. But the industrial scale of leaking made possible through the digitisation of information and the ability to communicate instantly across the globe – that *is* new. If it is to be revolutionary, however, we need a model for a new type of politics.

Conclusion

A Brave New World

*'The 21st century will be the century of the common
people. The century of you, of us.'*
Birgitta Jónsdóttir

The Chaos Computer Club was so named not because it set out
to cause chaos but rather because one of the founders, Wau
Holland, felt chaos theory offered the best explanation for how
the world actually worked. Rop Gonggrijp says the club is about
'adapting to a world which is (and always has been) much more
chaotic and non-deterministic than is often believed.' On
December 27 I go to hear Rop's keynote address to more than
2,000 international hackers and information activists at the
Club's annual conference in Berlin. Daniel Domscheit-Berg and
David House are here. Jacob Appelbaum decided at the last
minute not to attend; he's being routinely stopped and interrogated whenever he arrives back into the US from abroad. Julian
Assange, who spoke with Daniel at the previous two conferences,
couldn't come even if he'd wanted to as he's confined to a
mansion in England as part of his bail conditions while fighting
extradition to Sweden on sex assault allegations.

We are entering uncharted territory, Rop tells the audience,
and the CCC knows better than most how digital technologies
are revolutionising the world:

'Most of today's politicians realise that nobody in their ministry or any of their expensive consultants can tell them what is going on anymore. They have a steering wheel in their hands without a clue what – if anything – it is connected to. Our leaders are reassuring us that the ship will certainly survive the growing storm. But on closer inspection they are either quietly pocketing the silverware or discreetly making their way to the lifeboats.'

For those used to controlling citizens' communication, the digital age is frightening. Suddenly a seemingly powerless individual can, through interactive global networks, effectively challenge powerful individuals and institutions. How people view this depends largely on which camp they fall into. Those in the establishment may see this new empowerment of individuals as dangerous and de-stabilising; it is a threat to 'national security' and as such the Internet must be controlled. It presents a future where they are no longer the all-powerful gatekeepers of information. Others view this time of unprecedented freedom of speech as a gateway to a transformed political arena, an enlightened interconnected global democracy.

What is beyond doubt is that information now flows globally and nation states will find it increasingly hard to legislate using local laws. In the same way multi-national corporations use a pick-and-mix approach to world laws to maximise profitability, citizens can now take the same approach with freedom of speech. If you disagree with English judges issuing injunctions gagging the press and public from speaking about certain matters, whether it's footballers' sex lives or Trafigura's toxic waste dumping, then the Internet provides a way to publish in another jurisdiction where speech is better protected. If Iceland's Modern Media Initiative takes off, we could see that country become a global publishing centre where speech is protected by a selection of the world's strongest freedom of expression laws.

Nation states are fighting back, though, as they seek to extend their national laws to the world's citizenry. Few Americans can be well-versed (or interested) in the arcane practice of prior restraint in English courts. Nonetheless, an English law firm issued a court order to American company Twitter in May 2011 demanding the user data of tweeters who named Ryan Giggs, the Manchester United footballer who had been granted an injunction by an English judge to stop all reporting of an alleged affair. Equally, few citizens of the world will be totally clued up about the American surveillance laws I covered in Chapter 6, yet all information stored on Facebook, Twitter, Google or other American companies is subject to those laws. Birgitta may be an Icelandic politician, and Rop living in Holland, but their personal account data from social networks including Twitter was demanded under US law in January 2011, and it is in American courts where they must make their case.

What is likely to happen is that nations, realising their impotence at enforcing national laws against the world's citizens online, will band together to police the Internet. Already this is happening with those most threatened by an interconnected citizenry (despotic regimes but also the military, police and intelligence/security services of democractic countries) ramping up scare stories about the dangers of the Internet in order to bring it under their control. But speech is not like crime, and an Internet sheriff would not be in society's overall best interest. Those seeking regulation do so because it is in *their* best interest. It will serve to better protect their elite position as gatekeepers of power and information. A better idea would be to continue the current worldwide discussion about the legitimacy of certain laws banning speech. We're at a unique time where the world's cultures and laws are roiling together online in a giant cybermarket of ideas. Support is developing around laws based on *consent*, with online rebellions targeting those laws or actions perceived as unjust because they benefit only the powerful or are enforced through fear rather than consent. At the same time netizens are challenging England's elitist privacy law, they're also contesting Saudi Arabia's strictures against women drivers. By censoring the Internet, citizens will lose their new-found ability to challenge unjust law and abuse of power.

Instead of re-engineering the Internet to fit around unpopular laws and unpopular leaders, we could re-engineer our political structures to mirror the Internet. Instead of putting our faith in state intervention to control the Internet for our protection, we trust in the good that comes when individuals can speak and come together freely.

Free speech is not the great danger for humanity. Concentration of power is. We learn this lesson over and over again, and yet seem compelled eternally to repeat it. Communism, colonialism, monarchy, state socialism, tyranny – all become enemies of the people because they offer their citizens not too many opportunities to communicate or associate, but too few. Power is the dynamic force that fuels politics and it is this, not speech, which needs to be constantly monitored, controlled and checked. We view crimes against humanity as aberrations, individuals gone wild, when we should be seeing them through the prism of power. Abuse happens when a culture values some people more than others and those exercising power are not accountable for their actions.

The reason the Internet (and here I'm also including mobile phone networks) is so revolutionary is that it connects people. Interconnected people create a feedback loop that is highly responsive to concerns and as a result, self-organising groups arise all the time on the Internet to check power. In 2010 the US Congress tried to pass a law giving the president power to turn off the entire Internet (a kill switch). It was scheduled for a vote just one week after it was introduced. Aaron Swartz (whom we met in the previous chapter) discovered the bill, put up a public petition and in a week had more than 200,000 signatories. He used this to pressure Congress into shelving the bill pending further debate. Similarly, WikiLeaks can be seen as a counter-balance to the concentration of power by the US Government and military. When the site was shut down through extra-judicial means, the online group Anonymous launched Operation Payback to target companies believed to have colluded with certain politicians who wanted WikiLeaks taken offline. But as Anonymous's power grew, there arose other online groups to check its power until it was itself 'hacked' in 2011. The Internet is remarkable for allowing such balancing forces to come into existence incredibly quickly.

This is where chaos theory comes in. Chaos theory is found in many sciences that study dynamic systems such as the weather. Within nature there exist innumerable interconnections which make it impossible to accurately predict the future because even the tiniest difference in the initial conditions will have far-reaching effects. This is often called the 'butterfly effect' based on a paper given by MIT meteorologist Edward Lorenz in 1972. In 'Predictability: Does the Flap of a Butterfly's Wings in Brazil set off a Tornado in Texas?', Lorenz posited that the flapping wing represented the infinitesimal difference in initial conditions that led to a chain of events that would ultimately result in a tornado. Had the butterfly not flapped its wings, the trajectory of the events might have been completely different. Chaos theory demands we re-think our desire for order and certainty and instead develop a deeper understanding of the way things actually work, including human behaviour.

State intervention is the alternative to letting individuals cooperate together on a free Internet to solve problems. Its appeal lies in its immediacy. Politicians like passing laws and regulations because this makes them look as though they're doing something. But while the benefits of passing legislation come immediately, problems almost inevitably arise further down the road in ways that are so gradual and indirect that there is little incentive *not* to regulate. We could look at CALEA in the US as an example, since at the time the FBI argued it was necessary to crack down on crime. Whether putting backdoors into all telephone networks did reduce crime is hard to say, but it's certainly been indispensible to the world's despots in cracking down on pro-democracy campaigners. That was probably not the intention of the FBI and other state agents when they lobbied for this law; nevertheless that is one of its results. Good intentions matter little when the end result is the concentration of power. When this occurs in the more secretive arms of a state (such as the police and intelligence agencies) our concern should be even greater as it is in this secret soil that power germinates into tyranny and abuse.

In the 1990s the cypherpunks fought for individuals to have the same rights to communicate privately as their government. They won the fight for encryption but the battle to communicate freely is now entering a crucial stage. The telegraph and telephone were once free forms of contact, but that freedom was sacrificed in the name of greater security to state regulation. The information war is about ensuring the Internet doesn't follow the same route.

Re-wiring politics for the Digital Age

Rather than seeing an unregulated Internet as a threat, I believe it offers a solution: it acts as a global alerting system that highlights concerns in an incredibly efficient way. Centralised governments, even those bristling with giant state databases, can no longer accurately know nor predict what is happening in the world, and they will know even less as more people come online. Before the Internet, closed societies could remain stable because it took time for the people to realise that the rhetoric spouted by a dictator or state-controlled media did not reflect reality. People are loathe to act alone if they feel they are the only ones experiencing a problem. In the digital age, if a concern is held widely, those people can instantly unite and by doing so form a powerful interest group.

On 17 December 2010, a 26-year-old Tunisian fruit vendor, Mohamed Bouazizi, set himself ablaze to protest against state officials confiscating his fruit cart, which he claimed was done because he refused to pay a bribe. His act set in motion a train of events, starting with a local protest against corruption that was filmed by someone on a mobile phone. That video was shared with others online. It ended up on Facebook where it was harvested by satellite news channel Al Jazeera and was then broadcast across the Arab world. Protests spread across Tunisia and, united in their common goal to overthrow a corrupt regime, the people began protesting loudly and publicly against the leadership of Zine el-Abidine Ben Ali and his family. Ben Ali claimed the protestors were not representative of the people and a danger to national security. In a world of controlled communications that would have been the 'truth' the world heard. Precisely because of the uncontrolled nature of networked communication, alternative 'truths' could be told

– ones that better reflected the reality for the majority of people in Tunisia. The national uprising gained momentum and on 14 January, Ben Ali, who had been in power since 1987, was forced to give up his position, fleeing to Saudi Arabia.

The street vendor became a symbol of the ordinary person's frustration with a society where state machinery was used to benefit not the majority but a small yet powerful elite. Frustration with corruption and brutal security forces was shared by many others in the Middle East and soon citizens of other countries began protesting. On 25 January, Tahrir Square in Egypt filled with citizens demanding Hosni Mubarak's resignation. After 18 days of mass protest, the crowds got their wish and Mubarak resigned. Protestors in Libya, Yemen, Bahrain and Syria rose up against their ruling elites with varying degrees of success.

Rule by fear is not restricted to despotic countries, though the threat of state violence is certainly of greater concern in repressive societies. The protestors in the Middle East in 2011 had every reason to be fearful of authorities who unleashed violence against their citizens. The Internet may not have stopped the violence but it did ensure it was witnessed globally, completely refuting the official line that no violence was taking place or that it was all the fault of protestors. Whether it's political protests in Tunisia, Egypt or Iceland the world's citizens are using global communication networks to challenge the powerful and move away from political systems built around intimidation and secrecy.

The US diplomatic cables, while not offering a completely authoritative account of the world's governments, do present a too-rarely seen insider's look at how the world actually works. What becomes clear from reading these is how poorly most political systems are built, allowing the ruling classes to raid public resources for personal gain and use the security services as a tool to maintain and expand their power. The public's interest quickly becomes the king's interest. Politics is meant to help us live cooperatively despite our innate desire to act in our own personal interest. It shouldn't surprise anyone that we are wired to win. We are competitive, as all creatures are, and strive to attain the best life has to offer be that power, perks or prestige. What is surprising is how vulnerable most political systems are to being exploited by basic human nature, since they lack robust mechanisms that would check the inevitable drive to acquire and then concentrate power. There will always be those who try to game the system and some who actively seek to cause harm through maliciousness or psychological compulsion.

As the robustness of such a network is entirely dependent on the free flow of information, so laws protecting this free flow must be strengthened and enforced. Instead of penalties for disclosure we must move to a legal system where there are penalties for withholding information. Exemptions for the disclosure of official information must be for two reasons only: where it is in the public interest (not the nation's interest) or to prevent actual (not imagined) harm. An independent appeals organisation would monitor and oversee official secrecy to ensure it is not abused for personal or political purposes.

Rule by fear or consent

The clash of cultures we are facing isn't between countries or races or even religions, rather it is between individualism and authoritarianism. Authoritarians believe life can and should be based on a simple set of rules whereby everything can be predicted, including human behaviour. This is where states' increasing hunger to hoover up as much detailed information about their citizens as possible comes from – the belief that if only they had a full data set on every single person all social problems would be eliminated. It is one of the motivations behind the inexorable rise of the surveillance state.

Authoritarians offer citizens a deal: if we hand over our freedom, they will guarantee certainty and safety. This might have been possible in a closed society with little interaction between people, but it is a false promise in a knowledge economy where citizens are interconnected. If the best chaos theorists can't model the weather beyond a week, how does the National Security Agency think it can predict which of us will turn into a terrorist? If our intelligence agencies persist in monopolising knowledge we will see continued intelligence failures.

Over the past year I've thought a lot about censorship, surveillance and regulation of the Internet. Is it necessary? Is it really so dangerous to allow individuals an ability to associate and communicate freely? Certainly there exists a criminal minority who take advantage of the freedom of the Internet, but no one is arguing that crimes shouldn't be prosecuted. This is about allowing the vast majority of people to communicate without state intervention. Despite all the dire warnings, the prophesies of doom and destruction that were foretold by the Pentagon, the US State Department, Hosni Mubarak, even English High Court Judge Eady, I look at the fallout from all that was published in 2010, all the breaches to establishment power that occurred through a networked citizenry – and the good clearly outweighs the bad. From the uprising in Iceland to the ousting of dictators in the Middle East, free speech has fundamentally changed the world for the good.

Why, then, are the world's governments intent on controlling and regulating the Internet? Free speech is most threatening to authoritarian systems such as autocracies, militaries, the police and security services. Security services in principle exist for our protection but that is so only when they are accountable to the public for their considerable power. We are seeing a push by these agencies to move beyond the rule of law, to be accountable to no one but themselves. National security is becoming the new word of God to which all must submit in blind obedience. The decisions made, the liberties eroded, the crimes committed in the name of national security cannot be challenged because the information on which they are based remains secret.

Who is going to save us now?

Why do we remain so susceptible to arguments of an interventionist state when the historical record offers countless examples of the dangers that result from concentration of power? In spite of what history shows, there exists in human nature a desire for the mythic messiah, the idealised parent, some being or entity that is all knowing and all-protecting, to whom we can abdicate responsibility for this life and leave it to their guiding hand.

We seek a saviour, someone to rescue us from the problems of the world. A saviour is the simple story, the easy option and that is why it is so compelling. You don't have to *do* anything except believe. There's no need to negotiate with other people, or figure out how to create a robust system within the bizarre and contradictory parameters of human nature. I must admit I fell prey to this when I first met Julian Assange. He was going to lead the way to a bold new age. Instead I learned that power when concentrated is dangerous no matter who holds it or for whatever good intention. The real revolution happens in our own minds, when we stop believing there is someone or some agency who has all the answers, who is infallible and will save us, and instead come to realise we have that ability within ourselves. We may be susceptible to cults of personality, but we can build a check against this into our political systems.

The world may be more complex and uncertain than we would like, but giving away our freedom for the false promises of protection is not a sustainable solution. We are defined not just by what we preach, but by what we practice. We cannot claim to be an enlightened democratic society if we live in breach of these values, without the rule of law, without reason, or the rigorous commitment to truth.

The Internet is not the Wild West. A free-flowing network of communications may be exactly what takes us to the next stage of human development. Some have called the Internet the nervous system of the planet. It becomes harder to dehumanise when all of us are on the same network. Like the right hand declaring war on the left hand. This could be the system by which we take the next leap of evolution and move from a divided to a united species, learning how to build robust political structures that enable global cooperation.

Berlin is a good place to think about such questions as it's long been a haven for philosophers. Walking along Unter den Linden after the first day of the CCC conference, I come to Humboldt University where there's a quote in the main lobby by former graduate Karl Marx about the limits of philosophy: 'Philosophers have only ever interpreted the world, when the point is to change it.'

We now have a technology that unites individuals in such a way that we can create the first global democracy. Hundreds of millions of people are climbing out of poverty and the Internet gives them access to the sort of information that was previously accessible only to elite scholars. They can join a worldwide conversation and come together in infinite permutations to check power anywhere it concentrates. The greatest achievement isn't in producing technology, but using it to re-define the boundaries of what is possible.

Afterword to the paperback edition

If people didn't know there was an information war raging when this book was first published in August 2011, that truth is more apparent now that authoritarian and democratic governments alike are seeking to control both the Internet and our private communications.

The stakes have certainly been raised as a result of events in 2011, most notably the Arab Spring. Even hardline Internet pessimists had to admit the crucial role technology played in keeping protestors connected, despite the efforts of those in power.

Thus has it ever been. The success of the first Enlightenment depended on the invention of the printing press. These presses led to cheap journals which led to coffee houses where people came together regardless of status to discuss radical new ideas about equality, empiricism and democracy. The ideas weren't a result of the technology but they were able to spread fast and far as a result. So the Internet may be the ultimate invention of our own 'Enlightenment', with technology leading the way to a new world order where all forms of ordained authority are challenged and power is exercised through networks of networks rather than top-down, centralised bureaucracies.

In the revolutions of Egypt and Tunisia we witnessed the potential for emancipation that comes when people can speak and connect with each other across borders of geography, status and power. But it is one thing to overthrow old power structures and another to figure out who or what will replace them.

To give just one example: Power politics didn't end in Egypt on 11 February 2011 when President Hosni Mubarak resigned. Despite free elections, the old guard, supported by the military, fought furiously to maintain control. The Supreme Council of the Armed Forces (SCAF), which took over after Mubarak resigned, promised to cede power once elections were held. But in June 2012, the Supreme Court (whose judges were appointed by the Mubarak regime) dissolved the first freely elected parliament and, for a time, SCAF resumed control. The election of a new President brought hope, but there was worry, too, about how well the Islamist Muslim Brotherhood would uphold revolutionary principles of democracy and equal rights, particularly for women. Similar power struggles are raging across the Middle East as pro-democracy activists set about the difficult task of creating a more just, accountable and democratic political system.

I believe this desire for greater equality and more accountability from those in or exercising power (whether financial or political) is a growing global trend, what I call the Information Enlightenment.

The financial crisis, along with increasing public awareness about the prevalence and costs of tax evasion and official corruption, is also fuelling demands for greater transparency. Today about 137 jurisdictions around the world have disclosure regulations on politicians and politically exposed people according to a joint report by the World Bank and Stolen Asset Recovery Initiative (StAR) team. London's mayoral election in 2012 took an interesting turn when all candidates agreed to publish their tax returns and Prime Minister David Cameron claimed to be 'very relaxed' about publishing his own. 'The time is coming,' he said, 'for politicians to be open about their personal finances.' When it came to social media he was less relaxed.

In August 2011 riots began in London and spread across local areas and into other cities. Though the riots were initially sparked by the police shooting of an unarmed black man, they quickly degenerated into sprees of looting and arson. That the rioters (along with everyone else) communicated via social media seemed to cause politicians and the police particular concern. In a reactionary speech to Parliament, David Cameron pledged to explore ways of shutting down Facebook, Twitter and BlackBerry Messenger, and he was even pushing the idea of an Internet 'kill switch', until it was pointed out that shutting off the Internet was exactly what the Mubarak regime had done in Egypt during the pro-democracy protests.

Power pushes back

The traditional gatekeepers of power have now seen the full revolutionary force of the Internet and digitising information. It is not just China fighting for control: many of the world's governments – particularly their intelligence agencies, police and military – are seeking to put this genie back in its bottle. The pushback is coming in three forms: state surveillance, control of content, and control of infrastructure.

Politicians around the world introduced a raft of laws after 2010 to put citizens under increasingly intrusive levels of state surveillance. Sadly, some of the worst laws are coming from those countries purporting to be the most democratic, namely Britain and the United States. If the supposed torch-bearers of

Enlightenment values fall under the spell of authoritarianism, we must worry for the citizens of truly authoritarian countries. The West is in danger of abdicating its values and becoming a place where the universal surveillance of citizens is legitimised.

A draft Communications Data Bill was submitted to the British Parliament in June 2012, nicknamed the 'snooper's charter' as it would allow the police and security services to harvest data on social networking sites, webmail, voice calls over the Internet, and gaming. Although logs of websites could be recorded, pages within sites might not be. There would be no judicial oversight, rather police and spies would self-certify their surveillance. Internet service providers would be mandated to record and store a much broader range of personal data about their users which would require expensive re-engineering of their sites. The government was promising a blank cheque for phone and Internet firms to do this. The estimated cost: £1.8 billion.

Meanwhile in the US in 2012, the proposed Cyber Intelligence Sharing and Protection Act (CISPA) would give government and corporations vast new powers to track and share data about Americans' Internet use. Additionally, the use of surveillance drones to spy on Americans was being introduced. The biggest spy centre of all is expected to open in September 2013. The Utah Data Center run by the National Security Agency will be more than five times the size of the US Capitol. Its purpose, security expert James Bamford says, will be to 'intercept, decipher, analyze, and store vast swaths of the world's communications as they zap down from satellites and zip through the underground and undersea cables of international, foreign, and domestic networks'.

CISPA came about after the defeat of earlier bills such as the Stop Online Piracy Act (SOPA) that sought to censor and shut down websites based on their content. These bills attracted widespread protest from citizens as well as companies like

Google and organisations such as Wikipedia. However, copyright is still the acceptable face of democratic censorship and remains a recurring theme.

The biggest battle of all is coming up: who controls the Internet's infrastructure. Initially created as a borderless network of nodes populated by academics, geeks and the US Department of Defense, the Internet has expanded beyond all expectation. Governance of the Internet was passed to private-sector, technical non-profit-making organisations in the 1990s, and it is largely this bureaucratic hands-off approach that is credited with the Internet's phenomenal success. So if it ain't broke, why fix it? Although the Net is not 'controlled' by any one government, the United States, as the first major investor and creator, holds influence. The Internet Corporation for Assigned Names and Numbers (ICANN) is a non-profit organisation that regulates domain names and their suffixes (such as .com and .org). Based in California, it reports to the US Department of Commerce, meaning that effectively the US Government oversees the domain name process.

The US is increasingly taking full advantage of its influential status, not just as the home of companies like Google, Facebook and Twitter but also claiming jurisdiction on all websites registered in the US.

Countries such as Russia, China, Brazil, India and Iran are now challenging United States hegemony of the Internet and even calling for the creation of a new governing body to oversee Internet policy. The fight is currently playing out in the renegotiation of the United Nations treaty on International Telecommunications Regulations which began in February 2012 and could be complete by 2013. Countries such as Russia and China are lobbying to have the International Telecommunication Union (ITU) expand its remit from governing international telephone, television and radio networks to include the Internet. They want full sovereignty over 'their' Internet, which would fundamentally change the structure of the borderless network we know today.

Where are they now?

At the time of going to press, Bradley Manning was still awaiting trial two years after his arrest. A series of legal arguments, ironically enough around access to information, meant the full court martial was delayed from September to November 2012 or possibly even January 2013. After a public outcry over his treatment in the Marine Corp Brig in Quantico, Virginia (which State Department spokesman Philip Crowley called 'ridiculous and counterproductive and stupid', prompting the official's resignation two days later), Manning was transferred to a medium-security jail in Fort Leavenworth, Kansas.

At his pre-trial hearing in December 2011, digital forensics investigators from the US Army's Computer Crime Investigative Unit testified they had found 100,000 State Department cables on a computer Manning had used between November 2009 and May 2010; 400,000 US military reports from Iraq and 91,000 from Afghanistan on his computer and storage devices. They also recovered more than a dozen pages of encrypted chat logs on Manning's computer hard drive, between Manning and someone believed to be Julian Assange.

Meanwhile, Julian Assange remained in the UK fighting against extradition to Sweden where he was wanted for questioning in connection to the accusations of rape and sexual assault made by the two young Swedish women who had volunteered to help WikiLeaks. Throughout 2011 and 2012, Assange used the WikiLeaks Twitter feed to traduce the Swedish justice system. In his unauthorised autobiography he claimed that the women's sexual assault allegations were part of a conspiracy to have him extradited to the US for his work at WikiLeaks. That it is easier for the US to extradite from the UK than from Sweden was not mentioned.

Rather than return to Sweden and account for his personal behaviour, Assange spent the next two years fighting against the extradition with the help of a series of high-powered lawyers. Eventually in June 2012, the appeals came to an end when the UK's Supreme Court ruled Assange must return to Sweden in July 2012 for questioning. It looked as though we might get some answers about what happened in this case, but then a few days after the ruling he jumped bail and sought asylum at the Ecuadorian Embassy.

Another mystery remains as to how the funds from rights deals secured by his lawyers at the height of his fame, and donations to his personal defence funds were spent (although his legal costs were no doubt substantial). The only amount Assange is known to have given to Manning's defence fund was $15,000 as reported in this book.

Given Assange's well-documented penchant for dissembling and story-telling, it is questionable whether WikiLeaks ever really existed in the form that Assange described. The technical details of the secure submissions system have never been revealed and no one has been able to replicate it. No more leaks came from submissions after those attributed to Manning, though in February 2012 Assange announced publication of the 'Global Intelligence Files' – a series of emails taken from US security consultancy Stratfor by hackers associated with Anonymous and LulzSec. However, the following month, the FBI revealed that the leader of LulzSec, Hector Xavier Monsegur, a 28-year-old unemployed father of two living in New York and going by the online name 'Sabu', had pleaded guilty to computer hacking and other crimes and had been working as an informant for the FBI since August 2011. Court records state that an FBI-owned computer was used by Monsegur to facilitate the Stratfor leaks, suggesting the FBI may now have records of discussions between Assange, Anonymous and other hackers.

The operational security of WikiLeaks came in for criticism when the entire tranche of unredacted US diplomatic cables were found online in August 2011 though they had been there since 2010. The *Guardian* had unwittingly published the password to decrypt the file in its own WikiLeaks book, the authors taking Assange at his word when he told them the password to open their own file was temporary. It wasn't, and as a result the files were available to anyone. Assange decided to publish the entire lot himself in September 2011, leaving the identity of many confidential sources exposed. As the leaks dried up, so too did partnerships with the mainstream media. Assange would only give interviews or information to publications that provided favourable coverage. So *Rolling Stone* magazine, for example, received the Stratfor leak in return for writing a sycophantic interview.

The Icelandic activists severed their ties with Julian Assange in 2010 but they continued to work in the area of freedom of information and digital rights.

Birgitta Jónsdóttir remains an MP in the Icelandic Parliament and is working to implement the various pieces of legislation of the Icelandic Modern Media Initiative along with digital rights activist Smári McCarthy. By the summer of 2012, a law on source protection had been passed along with constitutional reform, and four different aspects of the original plan were pending ratification. Another four were in development. The project also received funding for its international promotion.

In November 2011, a US Federal Judge ruled that Twitter must hand over data from Jónsdóttir, the Dutch entrepreneur Rop Gonggrijp and hacker Jacob Appelbaum. The three appealed the ruling in January 2012. Jónsdóttir has not travelled to the US for more than a year under advice from the Icelandic State Department.

Smári McCarthy continues to travel extensively to meet with activists and politicians. In June 2012 he also joined the list of former WikiLeaks volunteers stopped and questioned by officials at the US border. He was asked about his affiliation with WikiLeaks and who he was meeting in the US. However, he was not detained in secret nor asked to work as an informant as claimed by Julian Assange.

Herbert Snorrason worked with former WikiLeaks second-in-command Daniel Domscheit-Berg on OpenLeaks but left in the winter of 2011 due to lack of funds. He is now a graduate student studying international relations at the University of Iceland and was spending the summer of 2012 creating and maintaining an open access site for the National University Library of Iceland. OpenLeaks, like all the other WikiLeaks-style sites, has failed to match in reality the rhetoric promised for an anonymous whistle-blowing portal.

Boston hacktivist Aaron Swartz showed himself to be unafraid of putting his radical principals about open access into action. In July 2011, he was arrested and charged with downloading 4.8 million academic articles between September 2010 and January 2011 from JSTOR, a research subscription service offering digitised copies of academic journals and documents. Swartz was accused of breaking into a computer wiring closet on MIT's campus and downloading the documents which prosecutors say he intended to share online. Swartz turned himself in and pleaded not guilty to charges including wire fraud, computer fraud and unlawfully obtaining information from a protected computer. He was released on a $100,000 unsecured bond and faces up to thirty-five years in prison, if convicted. His case was due to be heard July 2012.

Hope for the future?

We should not be surprised that those with power do not willingly give it up. Again it's worth looking at the first Enlightenment. Before 1650, Western civilisation was based on a core of faith, tradition and God-given (usually male) authority. No one questioned or challenged what Jonathan Israel described in his book *Radical Enlightenment* as 'the divinely ordained system of aristocracy, monarchy, land ownership and ecclesiastical authority.' After 1650, everything was questioned. The secularising of ideas challenged unjust systems that kept the masses powerless. People may have been frightened by the upheaval these new ideas of equality wrought, and the early Enlightenment philosophers could only promise that the long-term benefits would be immense.

We are all aware of the advantages of the first Enlightenment now: the revolution in science and medicine, the greatest number ever to be pulled out of poverty, the importance placed on human rights for all. And we would do well to think on this and remember the doomsayers of the Dark Ages Church who threatened chaos and that nothing good would ever come from allowing citizens to communicate freely with each other.

The currency of power is information, and the reason states push back is because they see power slipping away due largely to the technology of the Internet. The rapid innovation of technology means the only thing we can be certain about is uncertainty. What don't change, however, are our fundamental human values: freedom, justice, happiness.

Heather Brooke
London, 2012

Acknowledgements

I was fortunate to be able to work again with the wonderful Drummond Moir at William Heinemann, whose advice, skill, support and perspective kept everything running smoothly. A writer couldn't ask for a better editor. Much gratitude also to my agent, Karolina Sutton at Curtis Brown, for her help formulating the project in its initial stages and her advice throughout.

Thank you to all the people who talked to me for the book, most of whom are named in the chapters. A special mention to Aaron Swartz and Benjamin Mako Hill for their hospitality in Boston and to Chris Soghoian for his guidance on the complex law surrounding Internet and telephone surveillance in the chapter 'Land of the Free?'. Any errors are my own. Also thanks to the staff of the Electronic Frontier Foundation in San Francisco for their knowledge and helpfulness, particularly Kevin Bankston, Peter Eckersley, Katitza Rodriguez and Seth Schoen. I'm also grateful to the founders for ensuring such an important organisation exists: Mitch Kapor, John Barlow and John Gilmore.

Finally to my husband, Vaci, for his love, generosity and independent thinking. A truly remarkable man.

Index

1984 (novel) 14
9/11 pager messages 8, 117
Acxiom 152, 155
Afghan war logs 5, 95, 126, 160, 162, 164, 175–80, 183, 213
Ágústsson, Bogi 39
Ahmadinejad, President Mahmoud 108
Al Jazeera 181–2, 185, 233
Al Qaeda 5
All the President's Men (film) 11
Allan, Richard 148
Alseth, Brian 152–4, 156
Al-Shamari, Major Ali Jadallah Husayn 194
Althingi (Icelandic parliament) 34, 52
Amazon 14, 153, 223
American Civil Liberties Union (ACLU) 156, 172
Andersdotter, Amelia 52
Animal Farm (novel) 14
'Anonymous' 130, 224, 231
AOL 136
Apache helicopter video *see* 'Collateral Murder' video
Apache-SSL 134
Appelbaum, Jacob 25, 26–30, 51, 101, 115, 227
Apple 30, 107, 132, 146, 223
Arms Export Control Act (US) 99, 101
Assange, Julian 27, 37, 40–4, 48–54, 57–67, 74–5, 76–80, 90, 91–2, 115, 123–9, 161, 162, 163–5, 168–71, 178–86, 193, 194, 195–8, 210–13, 221–4, 227, 237
   appearance 64–5
   parentage 65–6
   rape and sexual assault allegations 183, 194, 224
Association for Computing Machinery (ACM) 18–20
Australia: censorship of the Internet 120
Aweys, Sheikh Hassan Dahir 170
Axten, Simon 147, 148

banking crisis, Iceland 38–9, 78

Bankston, Kevin 113, 118, 146
Barlow, John Perry 48, 105
Barron, Peter 141
Bates, Major Paddy Roy 45–6
Ben Ali, Zine el-Abidine 233–4
Berman, Jerry 113
Bernstein, Carl 11
Bidzos, Jim 100
BitTorrent 46
Boston University Information Lab
    & Design Space (BUILDS) 16, 20,
    21, 32
Boston University, Massachusetts 18
    student ID cards 18–20
Bouazizi, Mohamed 233
Boyes, Roger 37
Boyle, Professor James
    84, 87
Brandeis, Louis 132
Bremmer, Ian 214
Brunner, John 44
Bureau of Investigative Journalism
    185, 188

CACI (data broker) 155
Carter-Ruck solicitors 42–3
Catalist 154
C-base 22
censorship of the Internet 63,
    119–22, 236–7
Channel 4 181–2
Chaos Computer Club (CCC) 22,
    167–8, 175, 227
chaos theory 231–2
China 95–6, 119, 218
    attack on Google 143
    censorship of the Internet 119
    Great Chinese Firewall 95–6
Chmagh, Saeed 81, 89
ChoicePoint 154, 155, 156
Cisco 30
Clark, Danny 93, 202–8
Clinton, Hillary 96, 176

CNN 181–2
coffee houses, English 22
Coghlan, Tom 177
'Collateral Murder' video 6, 78–82,
    83, 89–92, 123, 125, 126, 178, 194,
    213
Combating Online Infringement and
    Counterfeits Act (COICA) (US)
    217
Communications Assistance for Law
    Enforcement Act (CALEA) (US)
    109, 113, 114, 118, 232
crackers 23
Creative Commons 87, 201
Crowley, P. J. 212–13
Crown copyright (UK) 85, 86, 87,
    88
Cryptonomicon (book) 45, 47, 51
Cybersecurity and Internet Freedom
    Act (US) 217
Cygnus Solutions 105
cypherpunks 98, 105, 106, 232

data brokering 150–60
data havens 44–7
Davies, Nick 73, 124, 125, 160–1,
    164, 165, 178–82, 193
Defense Advanced Research Projects
    Agency (DARPA) 103
Der Spiegel 160, 162, 180, 195, 197,
    219
Digital Collection System Network
    (DCSNet) 111
digital data 10–15
Digital Freedom Society, Iceland 40
digitisation of information 11, 12,
    105
Domscheit-Berg, Daniel 40–2, 48–54,
    165, 168, 169, 175, 183–4, 227
Dreyfus, Suelette 185
Dubois, Philip 99, 100
Dun and Bradstreet (data broker)
    155

Dynadot (ISP) 171, 173

Egypt, protests in 71, 110, 214, 234
Electronic Frontier Foundation (EFF)
    30, 48, 105, 113, 114, 171, 172
Ellsberg, Daniel 8, 12, 124, 191
encryption 97–102, 106, 118
English libel law 49
Espionage Act (US) 162
European Data Protection Directive
    157, 159
EveryDNS 223
Experian (data broker) 155
Expressen 183

Facebook 23, 131, 132, 139–40,
    147–9
FBI 111, 115, 232
Federal Communications
    Commission (FCC) 114
Feynman, Richard 210
First Amendment (US) 85, 86, 101
First Amendment Coalition 171, 172
First Data Resources 154
Fisher, Ian 211
Flat Earth News (book) 73, 124
Frayman, Harold 161
Free Culture (book) 88
Free Software Foundation 202
Furuly, Jan Gunnar 56–7, 67, 74
Future Shock (book) 71

Gates, Bill 31
Gates, Robert 90
Gibbs, Robert 176
Gibson, William 44
Giggs, Ryan 229
Gilmore, John 30, 105–6, 135, 159
GNU Project 23
Goetz, John 165
Goldman, Emma 28
Gonggrijp, Rop 51, 76–80, 115, 168,
    227, 229

Google 118, 119, 132, 134, 137–8,
    139, 140–7, 148
Googleplex 140
Googlesharing 142
government information 82–9
Graffiti Research Lab 21
Great Chinese Firewall 95–6
Grey, Stephen 181
Guantánamo files 126, 170
Guardian newspaper 42–3, 74, 125–6,
    160, 162, 180, 192, 165, 197–8,
    211, 218–19, 221–2, 225

HACK (Hungarian Autonomous
    Center for Knowledge) 22
hackers 23–24, 29, 30, 31, 44, 51, 143
hackerspaces 21–31
Hakkavélin (hackerspace) 40
Hansard 86
HavenCo Limited 45
Hicks, Brian 102
Hill, Benjamin Mako 201–4
Holland, Wau 227
House, Anthony 141, 146
House, David 16–21, 31–3, 101, 200,
    202, 207–9, 227
Hrafnsson, Kristinn 80–2, 185, 194,
    195

Ibrahim, Anwar 64
Iceland 34, 44. 47, 48, 121
    banking crisis 38–9, 78
Icelandic Modern Media Initiative
    (IMMI) 35, 54, 78, 121–2, 228
InfoUSA 154, 155
Ingason, Ingi Ragnar 80–2, 185, 212
injunctions (UK) 126, 160, 162, 171,
    171, 174, 228
    super injunctions 43, 132, 229
Intelligence Support Systems for
    Lawful Interception, Criminal
    Investigations and Intelligence
    Gathering 111–13

Internet Corporation for Assigned Names and Numbers (ICANN) 172
Internet interception 113–15
Internet Watch Foundation 120
Iran 107
Iran–Contra Affair 103
Iraq Body Count organisation 194
Iraq war 1–7, 74, 78–82, 83, 90
Iraq war logs 7, 95, 126, 160, 183, 185–6, 188, 193–4, 212, 220

J Curve, The (book) 214
J Curve theory of political revolutions 214–15
Jefferson, Thomas 84
Joint Worldwide Intelligence Communication System (JWICS) 3, 10
Jónsdóttir, Birgitta 35–6, 38–9, 40, 48–50, 52, 54, 78–80, 91, 115, 121–2, 165, 183–4, 229
journalism 68–9, 70–4
Julius Baer Bank 169, 170, 172–4, 223

Kapor, Mitch 105
Karn, Phil 101
Katz, Ian 162, 195, 210, 212, 219
Kaupthing Bank 39, 40, 42, 80, 121
Keller, Bill 160. 198, 221
Kerry, John 176
Khatchadourian, Raffi 79, 80
Khomeini, Ayatollah Ruhollah 107
Korolova, Aleksandra 149

Lackey, Ryan 46
Lamo, Adrian 93–5, 199, 206–9
Laurie, Ben 134–40, 159
Leigh, David 67, 74–5, 78, 126, 160–2, 165, 180, 187–9, 191, 193, 196–7, 210, 211, 220, 221
Lessig, Lawrence 87

Libya, protests in 110, 214, 234
Lieberman, Joe 223
Linux kernel 23
Lloyd's of London 23
Locke, John 84
lockpicking 32
Loosemore, Tom 86
Lorenz, Edward 231

Madison, James 84
Maliki, Iraqi Prime Minister 4
Manning, Bradley 93–5, 123, 165, 186, 193, 199–201, 204–10, 224
Marlinspike, Moxie 101, 142, 145, 159
Mascolo, Georg 197
Massachusetts Institute of Technology (MIT) 17, 200, 205
Mathews, Daniel 170, 171, 175–6
McCarthy, Smári 40, 42, 47–9, 50–2, 121, 184
McCord, Ethan 79, 83, 90
Meltdown Iceland (book) 37
Microsoft 30, 31, 132
Minton Report 43
mobile phone tracking 107
Moglen, Eben 48
Mohseni-Ejehei, Gholam Hossein 107
Moore, Gordon E. 12
Moore's Law 12
Mubarak, Hosni 234
Mullen, Admiral Mike 178
Mutashar, Sajad and Doaha 90

national security 213–18
Netflix 137, 158
New York Times 125–6, 136, 160, 194, 195, 198, 211, 219, 221–2
newspapers see journalism
NHS Spine 151
Noisebridge 22, 25, 26–9, 142
Nokia Siemens Networks 108

Norway 68
    newspapers 68–9
NTK (*Need to Know*) newsletter 31

O'Brien, Danny 31, 44–5, 131, 195
Obama, Barack 95
Oddsson, Davíd 39
Official Secrets Act (UK) 162
online privacy 130–3, 134–40, 142
Open Net Initiative 119
Open Organisation of Lockpickers
    21
OpenSSL 134
Operation Payback 224, 231
Orwell, George 14, 89

Patriot Act (US) 115, 116, 137, 151
'Pen register' court orders 110–13
Pentagon Papers 8, 12
PGP (Pretty Good Privacy) 97–102
Piratbyrån (the Piracy Bureau) 46
Pirate Bay 46–7
pirate radio 45
Poindexter, Admiral John 102, 103,
    143
Poulson, Kevin 94
Prévost, Antoine François
    22
privacy online 130–3, 134–40, 142,
    158
Privacy Rights Clearinghouse 155
Protecting Cyberspace as a National
    Asset Act (US) 217
Proudhon, Pierre-Joseph 41
Public Citizen Litigation Group 172

Really Simple Syndication (RSS 1.0)
    201
Reddit.com 201, 202
Reed Elsevier 155
Reuters 78, 91
Riegle, Robert 177
Robinson, Jennifer 195

RSA 100
Rubin, Scott 119
Rusbridger, Alan 126, 160, 162,
    195–8, 211, 213, 220, 222, 223
Russia 8, 216, 218

Safire, William 104
Saharkhiz, Isa 107–9
Saharkhiz, Mehdi 108
Saudi Arabia 30, 118, 189
Schmidt, Eric 12, 141
Schmitt, Daniel *see* Domscheit-Berg,
    Daniel
Schmitt, Eric 160
Science Applications International
    Corporation (SAIC) 102
Sealand 45–6
Shipton, John 170, 172, 173
Shockwave Rider, The (novel) 44
Siprnet (Secret Internet Protocol
    Router Network) 3, 7, 10
SKUP 56
Snorrason, Herbert 40–2, 47–9, 121,
    184
Software Freedom Law Center 48
Soghoian, Christopher 112
Spectator 23
Sprint (ISP) 111, 112
Sprout 22
Stephens, Mark 195–8
Stephenson, Neal 45, 51
Sterling, Bruce 44
Sterzer, Robin 98
Stone, Jim 18–21
'sudo leadership' 29
Sun Microsystems 30, 105
super injunctions (UK) 43, 132, 229
Swartz, Aaron 201–3, 231

Tableau 223
Task Force 373 ('kill or capture'
    squad) 164, 165, 176, 178
Taylor, Paul 111, 112

Terrorism Information Awareness 104
Text Secure 142
theyworkforyou.com 86
Times, The 177
Toffler, Alvin 71
Tor software 27
Total Information Awareness 102–5, 117
Trafigura 42–3, 169, 228
Traynor, Ian 125
Tunisia, protests in 110, 214, 225, 233–4
Twitter 43, 229

United Arab Emirates 118
United Nations 216
United States military 1–7
Unrepresented Nations and Peoples Organisation (UNPO) 215
unwarranted search and seizure 115–19
US Constitution 84
US diplomatic cables 126, 185–92 publication of 218–26
USCybercom 217

Vietnam war 8
VoteBuilder 154

Wahhabism 30
Wales, Jimmy 134
Wall Street Journal 133
Watkins, Tyler 93–5, 200, 202, 206
What is Property? 41
Whirlpool 119
Whisper Systems 142
Whole Earth 'Lectronic Link (WELL. com) 105
WikiLeaks 8, 9, 13, 27, 39–44, 91, 94, 49, 50, 52, 58, 62, 77, 95, 117, 119, 136, 163, 169–77, 182–5, 211, 213, 223–6, 231
court case against 171–3
first major press conference 79
legal threats against 58
ownership of 169–70, 175
registration of 170
setting up 77
Wikipedia 134
Wiretappers' Ball 111
wiretaps 110, 111
Woodward, Bob 11

Yoo, Amber 155

Zimmermann, Phil 97–102
Zuckerberg, Mark 131, 147, 148